Wittgenstein and the Cognitive Science of Religion

Also available from Bloomsbury

Arguing from Cognitive Science of Religion, by Hans van Eyghen
Philosophical Foundations of the Cognitive Science of Religion, by Robert N. McCauley with E. Thomas Lawson
The Creation of Wittgenstein, edited by Thomas H. Wallgren
Wittgenstein, Religion and Ethics, edited by Mikel Burley

Wittgenstein and the Cognitive Science of Religion

Interpreting Human Nature and the Mind

Edited by
Robert Vinten

BLOOMSBURY ACADEMIC
LONDON • NEW YORK • OXFORD • NEW DELHI • SYDNEY

BLOOMSBURY ACADEMIC
Bloomsbury Publishing Plc
50 Bedford Square, London, WC1B 3DP, UK
1385 Broadway, New York, NY 10018, USA
29 Earlsfort Terrace, Dublin 2, Ireland

BLOOMSBURY, BLOOMSBURY ACADEMIC and the Diana logo are trademarks of Bloomsbury Publishing Plc

First published in Great Britain 2023

A catalogue record for this book is available from the British Library.

A catalog record for this book is available from the Library of Congress.

ISBN: HB: 978-1-3503-2935-5
ePDF: 978-1-3503-2936-2
eBook: 978-1-3503-2937-9

Typeset by Deanta Global Publishing Services, Chennai, India

This book is dedicated to Gabriela Ferreira,
the most incredible of the incredibles.

Contents

Figures

Contributors

Mark Addis is Associate Dean – Knowledge Exchange at the Open University and a research associate at the Centre for Philosophy of Natural and Social Science at the London School of Economics. His publications include *Wittgenstein: A Guide for the Perplexed* (2006), *Wittgenstein: Making Sense of Other Minds* (1999) and the co-edited (with Robert L. Arrington) *Wittgenstein and Philosophy of Religion* (2001).

Guy Axtell is Professor in the Philosophy and Religious Studies department, Radford University, Virginia. He writes in philosophy of science, philosophy of religion and epistemology, and is a William James/pragmatism scholar. He has authored several books including *Problems of Religious Luck: Assessing the Limits of Reasonable Religious Disagreement* (2020).

Thomas Carroll is Senior Lecturer in Philosophy at the Chinese University of Hong Kong (Shenzhen). His areas of specialization are comparative ethics, history of analytic philosophy and philosophy of religion. He is the author of the book *Wittgenstein within the Philosophy of Religion* (2014) and of articles that have appeared in *Dao*, *Metaphilosophy*, *Philosophy East and West*, *Sophia* and *Religious Studies*. With Anne Blackburn, he prepared for publication John Clayton's posthumous book *Religions, Reasons, and Gods: Essays in Cross-Cultural Philosophy of Religion* (2006).

Florian Franken Figueiredo is a postdoctoral researcher at the Institute of Philosophy at NOVA University in Lisbon (IFILNOVA). He has authored a book and numerous journal articles about Wittgenstein's philosophy. He has recently been closely examining Wittgenstein's manuscripts from 1929 and is the editor of *Wittgenstein's Philosophy in 1929* (2023).

Casper Hesp is a PhD candidate in social neuroscience at the Faculty of Social and Behavioural Sciences, University of Amsterdam. His research aims to model and understand how social life develops in individuals and over generations.

Inês Hipólito is a postdoctoral researcher and lecturer in the Department of Philosophy at Humboldt-University in Berlin. She is also associated with the Berlin School of Mind and Brain and affiliated to the Theoretical Neurobiology Group, Wellcome Centre for Human Neuroimaging, University College London.

Christopher Hoyt is Associate Professor in the Department of Philosophy and Religion at Western Carolina University, where he has taught since 2003.

Rita McNamara is Senior Lecturer in Cross-Cultural Psychology at the Victoria University of Wellington and is the author of numerous articles on psychology, religion and cognitive science.

Olympia Panagiotidou is a postdoc researcher at the Department of the Study of Religion of the Aristotle University of Thessaloniki, Greece. She earned her PhD and holds a MA in Cognitive Science and the Study of Religion from Aristotle University and Aarhus University. She also holds a BA in History and Archaeology from Aristotle University. She is the author of the book *Healing, Disease and Placebo in Graeco-Roman Asclepius Temples: A Neurocognitive Approach* (2022), and the main author with Roger Beck of the book *The Roman Mithras Cult: A Cognitive Approach* (2017).

Roger Trigg is Senior Research Fellow, Ian Ramsey Centre, University of Oxford, and Emeritus Professor of Philosophy, University of Warwick. His particular research interests include the philosophy of religion and the philosophy of social science. His latest publication is *Faith* 2022.

Hans van Eyghen is Assistant Professor at Tilburg University, The Netherlands. His research focuses on epistemology and philosophy of religion. He has written on debunking arguments (e.g. *Arguing From Cognitive Science of Religion*, 2020) and cognitive science of religion (e.g. *Revising Cognitive and Evolutionary Sciences of Religion*, 2021).

Robert Vinten is currently a postdoctoral research fellow within the project Epistemology of Religious Belief: Wittgenstein, Grammar and the Contemporary World at Universidade Nova in Lisbon. He is the author of *Wittgenstein and the Social Sciences* (2020) and of numerous journal articles about Wittgenstein's philosophy.

Acknowledgements

I am grateful to the Fundação para a Ciência e a Tecnologia (FCT), the Portuguese funding agency, for their funding for the project Epistemology of Religious Belief: Wittgenstein, Grammar and the Contemporary World (PTDC/FER-FIL/32203/2017), without which this book would not have been possible. Their funding supported the workshops and lectures on Wittgenstein, naturalism and cognitive science of religion, where many of the papers that became the chapters of this book were first presented, as well as providing me with support as a postdoctoral researcher within that project. I would like to thank the principal investigators within that project, Nuno Venturinha (PI) and Sofia Miguens (Co-PI), for organizing the second and third ERB workshops with me and for their support in the early stages of putting together this book. I am grateful to Alexandra Dias-Fortes, Modesto Gómez-Alonso, Thomas Carroll, Gorazd Andrejč, Guy Axtell, Duncan Pritchard, Roger Trigg, Florian Franken Figueiredo, Olympia Panagiotidou, Hans van Eyghen and Rita McNamara, who were involved in those workshops both in giving presentations and in discussing the papers that were presented. Thanks are also due to E. Thomas Lawson, who gave the third ERB lecture within the project in February of 2021. He not only gave a wonderful lecture but also read and commented on several of the papers from the workshops held in 2020, which became chapters of this book. I'm grateful to him for taking the time to discuss some of the issues in the book with me via email.

Beyond the world of academia I am hugely indebted to my family for their support during work on this book. Anna Szpakowska, Janet Szpakowski, Michael Szpakowski, David Vinten and Jack Vinten have all been encouraging and supportive throughout the writing of this book and in the forty or so years of my life leading up to it. Lastly, I would like to thank Gabriela Ferreira for her support and for the concrete help she gave me in preparing the bibliography and formatting the chapters. This book is dedicated to her.

Introduction

Robert Vinten

The idea behind this volume is to bring together people working in various different disciplines in the hope of gaining a greater understanding of religious phenomena. A problem that it aims to overcome is that efforts to understand religions often go on quite independently of one another. One reason for this is that attempts to understand, interpret or explain religious phenomena go on in different disciplines (religious studies, psychology, philosophy, anthropology, neurophysiology, history) and people working in a particular discipline most often interact with other people working in the same discipline. Universities are organized into departments whose members more often interact with each other than with members of other departments. Academic conferences, more often than not, are attended by people working in a single discipline. Bringing people together from different backgrounds promises to open them up to new perspectives in a way that is mutually enriching of their work, offering new insights to them and to the audiences for their work. The contributors to this volume include academics working in psychology, cognitive science, neuroscience, history, religious studies and philosophy.

In particular, the volume brings together academics who have an interest in using the tools of cognitive science to understand religious phenomena, on the one hand, and academics who are inspired by the philosopher Ludwig Wittgenstein (1889–1951) in their endeavours to understand religion, on the other. I take it that the hope for mutual inspiration and insight through interdisciplinary discussion is in tune with the dominant spirit within cognitive science of religion (CSR), which sees itself as an interdisciplinary effort to understand (and to explain) religion. The founders of the cognitive science of religion, E. Thomas Lawson and Robert N. McCauley, make clear that their work draws on research in philosophy, religious studies, anthropology, psychology and linguistics.[1] McCauley himself works as a professor of philosophy at Emory University and Lawson is a professor emeritus of comparative religion at Western Michigan University, and they have worked together, collaboratively over many years. Pascal Boyer, another major figure within the field, with a background in anthropology and psychology, claims that insights are to be gained in religion by drawing on 'findings and models in cognitive psychology, anthropology, linguistics, and evolutionary biology'[2] and he also draws upon studies in neuro-psychology concerning brain activation in response to various stimuli.

Ludwig Wittgenstein is primarily known as a philosopher but he began his academic career as an aeronautical engineer. He went to Manchester in 1908 and conducted experiments with kites, designed a jet engine and worked on designing propellers.[3] After leaving Manchester he became captivated by philosophical questions in mathematics and logic and went to Cambridge to work on philosophy with Bertrand Russell. Writing about his influences in 1931, he lists 'Boltzmann, Hertz, Schopenhauer, Frege, Russell, Kraus, Loos, Weininger, Spengler, Sraffa'.[4] It is notable that this list includes only a few philosophers alongside scientists, cultural critics, a playwright, an architect, a historian and an economist. Wittgenstein took inspiration from work in the natural sciences, cultural criticism, architecture and history, and he also took a great interest in psychology. He was fascinated by the work of the psychologist and philosopher William James, commenting in a letter to Bertrand Russell in 1912 that 'Whenever I have time I now read James's "Varieties of religious exp[erience]". This book does me a *lot* of good.'[5] In his early days in Cambridge Wittgenstein conducted studies in the Psychological Laboratory there, investigating rhythm in music. He took a serious interest in Freud's work throughout much of his adult life and in conversations with the philosopher Rush Rhees, described himself as '"a disciple of Freud" and "a follower of Freud"' (although he was highly critical of Freud's thought and thought that the influence of psychoanalysis was harmful).[6] In thinking about religion and ritual Wittgenstein also took a keen interest in the work of the anthropologist J. G. Frazer and wrote a critique of Frazer's *The Golden Bough*. So, although Wittgenstein was, first and foremost, a philosopher, he was also something of a polymath who undertook the study of work in other disciplines and reflected on the relationships between those disciplines in his philosophical work.

I.1 Why Wittgenstein and cognitive science of religion?

We have seen that Wittgenstein (as well as many Wittgensteinians inspired by his work) and cognitive scientists all take a serious interest in religious phenomena that is informed by the study of work in several different disciplines. Moreover, Wittgenstein, Wittgensteinians and cognitive scientists of religion have engaged in investigations concerned with many of the same sorts of questions. There is the question, briefly mentioned earlier, of how the various disciplines involved in the study of religion relate to one another. One direction that this question might take us in is to ask questions about whether one of the disciplines (or regions of language) concerned is *reducible* to one of the other disciplines (or regions of language) involved in the study of religion (reductionism is discussed in Chapters 1, 2, 4, 5, 7, 8 and 11 of this book).[7] Another question connected with the relationships between the various disciplines is the question of *scientism*, which Hans-Johann Glock, in his *Wittgenstein Dictionary* describes as 'the imperialist tendencies of scientific thinking which result from the idea that science is the measure of all things'.[8] Wittgenstein, Wittgensteinians and cognitive scientists of religion have all expressed opposition to scientism but there are disagreements in their accounts of what exactly scientism amounts to[9] (scientism is discussed in Chapters 1, 5 and 7). The cognitive science of religion grew up in the 1990s, in part as a response to extreme forms of cultural *relativism* in the study of religion.[10] Wittgenstein's work on religion, as well as his work on other topics in philosophy, has also raised worries about questions of relativism (Chapters 1 and 2 contain discussions of relativism).[11]

Another common concern of Wittgensteinian philosophers and cognitive scientists of religion is the question of whether religion is a *natural* phenomenon (naturalism is the central theme of Chapter 7). Wittgenstein's work involved careful examinations of thought, belief, knowledge, faith and certainty, as well as discussions of conceptions of mind. These are obviously notions and conceptions that are central to cognitive science of religion and a proper grasp of them is essential for thinking about cognition in relation to religion (Chapter 10 gives an overview of various conceptions of mind in psychology, philosophy and religion).

However, despite these common concerns there has been little interaction between those who have taken inspiration from Wittgenstein in their efforts to understand religions and those working in cognitive science of religion. Studies from cognitive science of religion are rarely mentioned in Wittgensteinian discussions and Wittgenstein is rarely mentioned in the work of cognitive science of religion. There is a large literature discussing Wittgenstein and religion,[12] and there has been some discussion of cognitive science in the work of Wittgensteinians;[13] however, Wittgensteinians have rarely engaged with cognitive science of religion, in particular.[14] Similarly, cognitive scientists, despite recognizing a philosophical aspect to their discipline, have paid little attention to the work of one of the greatest philosophers of the twentieth century.[15]

That Wittgensteinians and cognitive scientists of religion rarely interact is clear but why should we think that bringing them into discussion would be productive? What do Wittgenstein and the Wittgensteinian philosophers inspired by him have to offer the cognitive science of religion? Why should Wittgensteinians pay attention to the studies and theories of cognitive science of religion? In the first place, the fact that the two groups have many overlapping concerns, as illustrated earlier, already provides some reason to think that interaction would be productive. Wittgenstein had much to say about reductionism and scientism, and he very carefully traced the relationships between concepts of knowledge, belief, certainty and faith. He took a great interest in William James's attempts to produce a 'science of religions'[16] and in James George Frazer's *The Golden Bough*, both of which can be seen as precursors to the modern cognitive science of religion.

Recent work by Wittgensteinians has developed Wittgenstein's insights into a critical engagement with contemporary cognitive scientists and neuroscientists. For example, Maxwell Bennett and Peter Hacker, in their *Philosophical Foundations of Neuroscience*, criticize Francis Crick, Colin Blakemore, Antonio Damasio, Daniel Dennett, J. P Frisby, Richard Gregory, Gerald Edelman and many others, for committing what they call 'the mereological fallacy'. The mereological fallacy is the 'mistake of ascribing to the constituent *parts* of an animal attributes that logically apply only to the animal as a whole'.[17] Bennett and Hacker are following in the footsteps of Wittgenstein here, who remarked that 'only of a living human being and what resembles (behaves like) a living human being can one say: it has sensations; it sees; is blind; hears; is deaf; is conscious or unconscious'.[18] This criticism, made of neuroscientists and cognitive scientists, obviously carries over to many of those working in the cognitive science of religion. It is commonplace for researchers in the cognitive science of religion to ascribe psychological attributes to parts of animals (i.e. their brains), which logically apply to the animal as a whole. In Chapter 3 of this book I suggest that this is a major problem for Boyer and in Chapter 4 Hans van Eyghen offers a response to these worries.

It is worth noting that even if van Eyghen is correct to claim that the mereological fallacy is not a serious worry for those working in the cognitive science of religion that his response to it differs from others who work in this area. Whereas van Eyghen claims that 'states like perceiving and inferring' can be attributed to brains,[19] when those states fulfil the relevant functional criteria, Daniel Dennett claims that the attribution of psychological states to 'sub-personal' systems is only appropriate analogically. In the *Philosophical Foundations of Neuroscience*, Bennett and Hacker survey a wide range of different kinds of responses to the charge of committing the mereological fallacy from neuroscientists.[20] There is clearly work to be done in the cognitive science of religion in figuring out exactly how to respond to claims that researchers in the area are misusing psychological expressions in attributing psychological attributes to brains. There is a serious *philosophical* problem here that they have to deal with, involving getting clear about the appropriate application of psychological expressions.

In addition to Wittgensteinian criticisms of contemporary cognitive scientists that can be carried over to the cognitive science of religion it is also useful to think about Wittgenstein's critical engagement with the precursors to cognitive science of religion such as James George Frazer and its relevance to critical engagement with contemporary cognitive science. Wittgenstein criticized Frazer's account of rituals and magic, which thought of them as a kind of proto-science, with the rituals being interpreted by Frazer as being conducted to bring about a certain desired end, and 'magic' being thought of as an attempt at explaining worldly phenomena in supernatural terms where a natural explanation was lacking. Wittgenstein pointed out that there are many familiar ritualistic actions that people engage in where there was no question of thinking of the ritual as being causally efficacious in bringing about a certain desired end. For example, someone might kiss a picture of a loved one and they would not obviously do so believing that their action will have an effect on the loved one.[21] Illumination in understanding unfamiliar rituals might come through comparing them to this kind of behaviour rather than thinking of them as being erroneous attempts to bring about an effect.

Christopher Hoyt's chapter in this volume looks at efforts to understand, interpret or explain rituals, including Pascal Boyer's attempts to explain burial rituals in terms of inference systems. He argues that Boyer makes similar kinds of mistakes to the ones Wittgenstein identified in Frazer's work: treating rituals as if they were the upshot of confusion resulting from inference systems operating below the level of our conscious thinking, much like Frazer treated rituals as a confused upshot of two principles of magical thinking. Boyer himself is critical of Frazer's intellectualism,[22] but if Hoyt's objections against Boyer's account of rituals are correct, then it seems that Boyer has not gone far enough in his rejection of intellectualism.

So, we have seen that cognitive scientists of religion have something to gain from looking at Wittgensteinian objections to their use of psychological predicates and that they might also gain from looking at Wittgensteinian objections to the precursors of cognitive science of religion. Even if they disagree with those objections, there is not a settled view about how best to respond to them and so the work of Wittgenstein and Wittgensteinians serves as a reminder of deep problems and unsettled questions. What about the reverse case? What do Wittgensteinians have to gain from engaging with cognitive science of religion?

Cognitive science of religion is a relatively young research area but it has grown in the past thirty or so years since its foundation and there are now research centres dedicated to developing research in the cognitive science of religion, such as the Religion, Cognition and Culture (RCC) research centre in Aarhus, and journals dedicated to discussing research, such as the *Journal for the Cognitive Science of Religion* (established in 2012). The fact that it is a thriving area of research that has been given a philosophical defence suggests that it is something philosophers should engage with. Although Wittgensteinian philosophers might well think that the conceptual frameworks used by people working in CSR involve confusions (such as the confusions about the application of psychological predicates to brains illustrated earlier), there is much in common both in terms of concerns and in terms of approaches between Wittgensteinian philosophers and cognitive scientists of religion. Both Wittgensteinians and cognitive scientists of religion claim to be anti-theoretical[23] in their attempts to comprehend religious practices. Although cognitive scientists very often talk in terms of theory they do not necessarily think that there can be a theory *of religion* as such.[24] Both Wittgensteinians and cognitive scientists of religion tend to think that no definition in terms of necessary and sufficient conditions, can be given of 'religion'. In Wittgensteinian terms, we might say that cognitive scientists of religion are open to the idea that 'religion' might be a 'family resemblance concept'. Wittgensteinians and cognitive scientists profess to be opposed to scientism, and in both cases this is combined with a recognition that there is a great variety of different kinds of explanation.[25]

Moreover, although Wittgenstein rejected Frazer's intellectualism, this did not imply that Wittgenstein saw no value in Frazer's work. In 'Remarks on Frazer's *Golden Bough*' Wittgenstein says that what one needs in order to understand the variety of ritualistic practices is a perspicuous representation, which allows one to see connections between various different practices.[26] What one has to do is to 'correctly piece together what one *knows*',[27] but in order to do this it is useful to have experience of or a description of a great variety of ritualistic practices, which is something that Frazer provided. I suspect that one reason that Wittgenstein was fascinated with James's *Varieties of Religious Experience* and Frazer's *Golden Bough* was that they provide descriptions of a great variety of cases of religious conversion (James) and ritual (Frazer). Piling case upon case and looking for similarities and dissimilarities between them is a way to gain an understanding of these aspects of religion, as well as a way for gaining an appreciation of the variety of them. This is just as true today as it was then. Wittgensteinian philosophers tend to focus on conceptual matters rather than experiences or descriptions of those experiences but attention to the kind of data anthropologists provide, properly understood, is essential to gaining an understanding of religious practices. In a similar way to which Wittgenstein found value in engaging with Frazer's work, Wittgensteinians (and others) today can find value in engaging with the work of cognitive scientists of religion.

I.2 An overview of the chapters

Eight of the eleven chapters in this volume were first presented at workshops at Universidade Nova in Lisbon, with other contributors being amongst the audiences for those presentations. The interdisciplinary spirit mentioned at the beginning of this introduction infuses the chapters not only in the sense that the authors of those chapters come from backgrounds in different disciplines but also in the sense that there was interaction in the development of the chapters. The first two chapters here started out as a paper presented by Roger Trigg with a response given by Florian Franken Figueiredo at the third Epistemology of Religious Belief (ERB) workshop, hosted by Universidade Nova in Lisbon in December of 2020 and organized by myself, Nuno Venturinha and Sofia Miguens. My own chapter, Chapter 3 of this volume, began as a presentation at the second ERB workshop (in July 2020) and Hans van Eyghen gave a response to my paper when he gave his own presentation at the third ERB workshop. The chapters from Olympia Panagiotidou, Thomas Carroll, Guy Axtell and Rita McNamara were also presented at these workshops, and in every case there was a discussion of the papers presented at the workshops involving the attendees. Christopher Hoyt's chapter began as a paper presented at an earlier conference in Seville, in 2019, at which I was present.[28] E. Thomas Lawson also presented work related to the ERB workshops at a later date, in February of 2021, and many of the contributors to this volume were present to discuss Tom Lawson's presentation. The chapters from Mark Addis and from Inês Hipólito and Casper Hesp were commissioned after the two ERB workshops.

Roger Trigg's chapter starts by situating Wittgenstein and cognitive science of religion within broader debates about the nature of social scientific understanding and reductionism. He suggests that Wittgenstein's early work, the *Tractatus Logico-Philosophicus*, presented a form of realism and that when Wittgenstein rejected his Tractarian views in his later work he generated problems about objectivity and truth. He worries that in Wittgenstein's work 'religion is . . . treated as a separate system of belief with its own standards, without appeal to anything external to it to confirm or falsify it'.[29] Considerations in Wittgenstein's later philosophy about differences between 'language games' or 'forms of life' 'of different conceptual systems, whether physics and oracles, science and religion, or anything else, inevitably becomes a type of conceptual relativism that regards such systems as insulated from each other'.[30] Cognitive science of religion, on the other hand, according to Trigg, does not present similar worries about relativism but faces its own problems in the form of problematic forms of reductionism. What we need to do, he thinks, is to steer a course between the two. We should acknowledge that Wittgenstein was correct to worry about scientism and reductionism in philosophy and the social sciences but suggests that in acknowledging those worries we must be careful not to go to the extreme of adopting a relativistic viewpoint.

Florian Franken Figueiredo, in the second chapter, agrees with Trigg that the correct approach is to find a middle ground between reductionism and scientism on the one hand, and social construction on the other but, *contra* Trigg, he thinks that Wittgenstein's ideas are particularly useful in finding this middle ground. His chapter begins by presenting us with a brief history of the development of the cognitive science of religion and then gives us an overview of basic presuppositions shared by proponents of CSR. His chapter then moves on to discuss Trigg's account of Wittgenstein's philosophical development. Figueiredo's own, alternative, account exhibits sensitivity to the way in which Wittgenstein's thought developed between the *Tractatus* and the *Philosophical Investigations* which is the product of research that he has undertaken into Wittgenstein's manuscripts from 1929 into the 1930s. Figueiredo rejects Trigg's understanding of Wittgenstein as holding the view that reality is socially constructed. Language-games, Figueiredo argues, '*are* part of reality',[31] and not what constructs it, on Wittgenstein's view. Wittgenstein, according to Figueiredo's account, did not think of language and reality as two separate realms, such that a theory is required to establish the relationship between the two. Wittgenstein's rejection of philosophical realism did not mean giving up on the idea of objective truth, just a reconceptualization of it, and so Figueiredo suggests that Wittgenstein, in his later work, accepted a 'modified notion of realism'.[32]

My own chapter takes Wittgenstein's remark that 'only of a living human being and what resembles (behaves like) a living human being can one say: it has sensations; it sees; is blind; hears; is deaf; is conscious or unconscious' and uses it to criticize Pascal Boyer's attempts to explain the acquisition and spread of religious ideas in terms of 'inference systems' in the brain. The criticism implicates more than just Boyer's work. Many other researchers in the cognitive science of religion talk in similar terms, attributing psychological states to parts of the brain.

In the fourth chapter Hans van Eyghen provides a defence of Boyer's attribution of psychological states to the brain. He argues that it is legitimate to speak of representations in the brain and of brains perceiving and inferring. These attributions to brains can be made, van Eyghen argues, if we define the relevant expressions in functional terms. If we define 'perception' according to its function, that is, taking in information of the environment, then it is clear, van Eyghen thinks, that this is something that both human beings and brains can do. Similarly, inferring, defined in functional terms, amounts to drawing conclusions from or responding to available information. Why should we prefer van Eyghen's functional definition of these terms to the account provided by Wittgenstein and Wittgensteinian philosophers? He argues that the definitions used by empirical scientists are preferable because those definitions from 'empirical scientists are closer to common sense than those of Hacker & Bennett since theirs build on complex argumentation'.[33] Even if we are reluctant to accept the functional definitions, the very fact that we can make a comparison between the functional definition and the Wittgensteinian account suggests that the functional definitions make sense. Moreover, van Eyghen claims that it is clear that talk of brains inferring and perceiving makes sense because there is a large literature in the cognitive science of religion where researchers write in this way and those researchers make themselves understood to one another: 'The wide discussion on cognitivist theories that rely on such attributions is good enough reason to argue that such attribution is sensical',[34] he says.

In the fifth chapter Christopher Hoyt provides another critical take on the work of Pascal Boyer. He argues that the causal, psychological explanations that cognitive scientists of religion provide are not well suited to explaining religion. Hoyt argues that Boyer's explanatory tools make him blind to religious meaning. They tend to result in accounts of religious beliefs and religious practices as being rooted in confused, unconscious beliefs – the work of inference systems operating below the level of consciousness. An alternative form of explanation is needed, Hoyt thinks, if we want to gain an understanding of religious practices and beliefs. He proposes that Wittgenstein's suggestion of providing 'perspicuous representations' is better suited to gaining an understanding of religion. He concludes that '[t]o understand the religious lives of others, we must strive to see the world as they do, to see how their solutions to the problems of life work for them'.[35]

Olympia Panagiotidou suggests that cognitive scientists of religion share a lot of common ground with Wittgenstein and Wittgensteinian philosophers and that they can use that as a basis for collaborative work as well as material for fruitful discussions in Chapter 6. She begins by tracing the development of Wittgenstein's views from the early *Tractatus Logico-Philosophicus* to his later work that was published posthumously as *Philosophical Investigations*. She notes that Wittgenstein's conception of philosophy, both early and late, distinguishes philosophy quite sharply from the natural sciences. Nonetheless, there is no incompatibility between philosophy and the natural sciences; rather philosophy complements natural science by investigating the use of concepts in the sciences, as well as in other areas of life, and by detecting misuses, misunderstandings and fallacies. She then goes on to look at some points of intersection between Wittgenstein's work, both early and late, and work in the cognitive science of religion. Having done this, Panagiotidou goes on to look at historical investigations, in particular, her own area of research, and details how she thinks that they can be situated in relation to investigations in both philosophy and natural science. She describes how her own historical research developed, adopting methods and findings from neurocognitive research and applying it to understanding religious systems of Graeco-Roman antiquity. She rejected brain–body dualism in favour of Armin Geertz's bio-cultural theory of religion which sees human beings as being embrained and embodied organisms situated in particular circumstances and interacting with each other as well as their environment. She concludes her chapter by suggesting that interdisciplinary work between historians, philosophers and natural sciences will enhance our understanding: 'by joining their methods and accomplishing their distinct roles, researchers from the disciplines of philosophy, history and cognitive sciences may deepen their knowledge and understanding of both the human beings and the world'.[36]

The focus of Thomas Carroll's chapter is the role of different forms of naturalism in both cognitive science of religion and the work of Ludwig Wittgenstein. He argues that Wittgenstein's remarks about religion 'exemplify a way of thinking about the naturalness of religious phenomena, and that naturalistic depiction is part of the clarificatory work of philosophy'[37] but he distinguishes the way of thinking about the naturalness of religion found in Wittgenstein's work from characterizations of religion as a natural (or supernatural) phenomenon found in the cognitive science of religion. If naturalism is defined in such a way that it basically amounts to a kind of scientism then Wittgenstein would no doubt reject it. This kind of reductive naturalism, Carroll argues, is found in the work of Daniel Dennett. It is used by Dennett as a means for rejecting religious claims, since they conflict with scientific ones, on his view. However, although Wittgenstein's views are in tension with strong forms of naturalism this does not mean that Wittgenstein cannot be considered a naturalist at all. We might consider Wittgenstein's work as involving a kind of non-reductive naturalism, Carroll suggests. In Wittgenstein's later work there are references to the natural history of human beings and there is an emphasis on keeping common human propensities in mind as a means to understand unfamiliar rituals.

Guy Axtell's chapter also looks at a way of thinking about religions as natural within the cognitive science of religion, namely Robert McCauley's claim that religion is natural and science is not. McCauley's claim is that human beings are naturally inclined to adopt and transmit religious beliefs, given the cognitive equipment they are endowed with. However, science does not come naturally to human beings in the same way. According to McCauley, 'human minds will regularly follow paths and introduce variations that more closely harmonize with their natural cognitive prejudices'.[38] However, whether a belief comes naturally to us or not is not a good reason in itself to accept or reject it. Axtell hopes to shed light in his chapter on beliefs that we *ought* to accept but that nonetheless do not come naturally to us given our cognitive biases by combining insights from cognitive scientists with observations from Wittgenstein's later work, particularly 'Lectures on Religious Belief'.

A concern with normativity, with 'oughts', is also central to Mark Addis's chapter, which examines normative cognition in religion and the treatment that it receives in the work of cognitive scientists of religion. Addis is critical of approaches to normative cognition in religion from cognitive scientists but this does not imply a rejection of cognitive science of religion altogether. Instead, he argues that '[t]he methodological use of Wittgenstein in the cognitive science of religion offers the potential to ground theories and research programs in a way which better integrates and aligns the language, practice, and theoretical discussion of religion'.[39] Addis emphasizes problematic reductionistic tendencies in the cognitive science of religion that are commented upon in other chapters of this book but he also deepens the criticism by noting that the conceptual frameworks of many of those working in the cognitive science of religion implicitly commit them to there being private mental representations and 'the commitment which reductionist approaches favouring methodological individualism have to private mental representations is a consequence of how these approaches account for mental representations acquiring and exercising normativity'.[40] This commitment to privacy can be challenged by the various remarks made by Wittgenstein that have been labelled the 'private language argument'.

In Chapter 10 Rita McNamara provides us with a useful reminder of both the diversity of religious beliefs and practices around the world and the diversity of conceptions of mind associated with them. Theories in the cognitive science of religion take the operations of the brain and inferences about the minds of others to be central to explaining social behaviour. However, McNamara suggests that more attention should be given to the cultural setting in which these views about minds and brains have arisen. Cognitive scientists of religion too often fail to take into account the world view which underlies and influences the theories they develop and the methods they use. She examines Western models of mind in both formal and 'folk' scientific varieties. She then goes on to look at models of the mind in non-Western traditions, including formal models of mind found in Advaita Vedanta, Buddhism and Samkhya, before examining alternative 'folk' conceptions of mind found in the Americas and in the Pacific. There are considerable differences between these various models in terms of the weight they give to intentions in understanding human behaviour, the weight they give to the influence of the situation humans find themselves in, and in terms of the extent to which the various models allow that we can have knowledge of other people's minds. Her chapter suggests that greater attention should be given to the presuppositions underlying conceptualizations of human psychology and an acknowledgement of the diversity of conceptions that are to be found.

Finally, in the eleventh chapter, Inês Hipólito and Casper Hesp try to unite insights from Wittgenstein, cognitive science and complex systems theory to present a communal perspective on religion. They begin their chapter by highlighting the diversity of religions in terms of (i) their understanding of the self (*self-relation*), (ii) their understanding of *salvation* or of how they express the yearning for some kind of resolution, (iii) their understanding of *devotion* to a larger whole, which is regarded as sacred and (iv) in terms of the way in which religions encourage *veneration* of highly dedicated devotees of the religion in question. They go on to look at how cognitive science might try to comprehend this diversity of religious practices and beliefs. They bring a number of objections against using the Modularity of Mind framework, before looking at predictive accounts, and finally at a Wittgensteinian alternative, particularly drawing on Wittgenstein's 'Remarks on Frazer's *Golden Bough*', and his later remarks from *Philosophical Investigations* and *On Certainty*. Inspired by Wittgenstein they hope to provide a perspicuous representation of religion in terms of many different layers of description, and their way of developing this multi-perspectival approach is to do so 'in light of multi-scale formulations of Markov blankets as developed by Friston'.[41]

There is a range of different approaches to understanding religion in the eleven chapters of this volume. There are disagreements about realism, relativism, reductionism, naturalism, and the role of the natural and social sciences (as well as philosophy and history). I hope that by bringing together a collection of new work discussing these issues that some light has been shed on them.

Notes

1 Lawson and McCauley's *Rethinking Religion: Connecting cognition and culture* is often credited with being the foundational text of the cognitive science of religion (see, for example, White 2021, 6). In that book they say their work is likely to be of interest to scholars in 'philosophy, religion, anthropology, psychology, and linguistics' and that 'We owe much to the work of researchers from each of those fields' (8). More recently McCauley has again emphasized the interdisciplinary nature of cognitive science of religion and how he thinks it is mutually enriching, in an attempt to allay worries that cognitive science of religion is trying to explain away religion (McCauley with Lawson 2017, 23).

2 2001, 2.

3 Monk 1991, 28-35.
4 1998, 16e.
5 McGuinness 2008, 30.
6 Wittgenstein 1966, 41.
7 An example of a fairly recent discussion of reductionism from a cognitive scientist of religion McCauley 2017.
8 Glock 1996, 341.
9 For example, in *The Blue Book* Wittgenstein says that 'Our craving for generality has another main source: our preoccupation with the method of science. I mean the method of reducing the explanation of natural phenomena to the smallest possible number of primitive natural laws . . . Philosophers constantly see the method of science before their eyes, and are irresistibly tempted to ask and answer questions in the way science does. This tendency is the real source of metaphysics, and leads the philosopher into complete darkness' (18). Robert McCauley, in describing the explanatory pluralism that he favours in the cognitive science of religion says that 'Explanatory pluralism stresses, first, that science is not the only game in town and that it is not the only way that we acquire knowledge (no scientism)' (2017, 3).
10 White 2021, 5–7.
11 Note: there is significant disagreement amongst Wittgensteinians about whether some form of relativism is present in Wittgenstein's work. Martin Kusch (2013) presents an overview of the disagreements. He is also the editor of the recent (2020) *Routledge Handbook of Philosophy of Relativism*, which provides an excellent overview of discussions of relativism in many different areas of philosophy. Another excellent recent (2020) survey of relativism is Annalisa Coliva and Maria Baghramian's *Relativism*. In Chapter 2 of Vinten 2020 I give a survey of interpretations of Wittgenstein's work in terms of relativism and also respond to accusations of cognitive relativism levelled at Wittgenstein by Roger Trigg.
12 See, for example, Malcolm 1993; Arrington and Addis 2001; Nielsen and Phillips 2005; Schönbaumsfeld 2007; Graham 2014; Mulhall 2015; Burley 2018; Andrejč and Weiss 2019. This is just a small selection from the enormous literature.
13 See, for example, Hutto and Myin 2013; Loughlin 2021; Bennett and Hacker 2022.
14 Some of the contributors to this volume are something of an exception to this rule. Guy Axtell's 2019 book *Problems of Religious Luck* discusses both Wittgenstein and cognitive science of religion (although Axtell himself is not a Wittgensteinian). Roger Trigg has written about both Wittgenstein's work and cognitive science of religion (although, again, Trigg is not a Wittgensteinian – see, for example Trigg 1990 and Trigg and Barrett 2014). Mark Addis has given a lecture in Bergen 'Methodology and Meaning in the Cognitive Science of Religion' 2009, which brought Wittgensteinian considerations to bear on cognitive science of religion. Christopher Hoyt gave a lecture at a conference in Seville in 2019 entitled 'The spurious inner presumed by the Cognitive Science of Religion', which he developed into a chapter for this book (Chapter 5).
15 In Lawson and McCauley's *Rethinking Religion* there is a brief, critical, engagement with the Wittgensteinian philosopher, Peter Winch (19–20) but Wittgenstein himself rarely features in the work of cognitive scientists of religion.
16 James 1982 [1902], 455.
17 2022, 84.
18 2009, §281.
19 66.

20 2022, 85–6.
21 1993b, 123.
22 See Boyer 2001, 15–16.
23 Boyer 1994, 32.
24 Pyysiäinen 2012, 6.
25 See, for example, Chapter 1 of McCauley 2017.
26 1993b, 133.
27 1993b, 121.
28 The conference was entitled 'Wittgenstein: Beyond the Inner-Outer Picture' and Christopher Hoyt's presentation had the title 'The spurious inner presumed by the Cognitive Science of Religion'.
29 17.
30 18.
31 28.
32 35.
33 64.
34 61.
35 89.
36 106.
37 110.
38 McCauley 2013, 238.
39 151.
40 155.
41 189.

1

Wittgenstein, concepts and human nature

Roger Trigg

1.1 The logical priority of conceptual thought

The philosophy of social science has for many years seen tension between those who view the social sciences on the model of the physical sciences and those who see them as pursuing a very different path. The basic distinction is between those who want to reduce social science ultimately to physics, pursuing a reductionist and naturalist policy, and those who see human sciences as being involved in something quite distinctive. The latter have been concerned above all with human consciousness and understanding. They are concerned not just with how people behave, or their physical movements, but how they understand their behaviour. The stress is put on human conceptual thinking and the logical priority of concepts. The point is often made that human beings do not just *behave* but *act*. Human actions are not arbitrary bodily movements but are endowed with a meaning, that has to be properly understood. Social science can be seen as concerned with actions that can only make sense in a social setting with socially agreed meanings and agreed understandings. It may be a conceptual enterprise in a way in which physics is allegedly not. We do not have to worry about what electrons think they are doing, whereas understanding any society must, it seems, involve an appreciation of the participants' view of the significance of their actions.

The views of Wittgenstein in his later work, just before his death in 1951, have been particularly influential in many branches of social science, such as anthropology. His views have meshed in with continental European stress on the relevance and importance of hermeneutics, the interpretation of meanings. The assumption has been that anthropologists cannot study a society from the security of their own armchairs. They cannot even land on a remote island and take for granted they understand what they are seeing. They must live with the community as participants. This emphasis on fieldwork in social anthropology stems from an understanding that it is only by participating in a particular society that anthropologists can properly understand the significance of what is happening. Social context is all important and a total outsider can never fully grasp what is happening. To take a lurid example, human sacrifice may be murder in some people's eyes, but it might be seen very differently given a particular religious context and a particular kind of society.

Wittgenstein's later philosophy is oriented to human beings and their activities in a way that contrasts with more rigidly scientific understandings of human activity. For example, that tended to be judged as a mere species of animal behaviour in the early years of what was called sociobiology[1]. That has now been absorbed into a more general evolutionary psychology. Certainly, though, there was an initial tendency to see behaviour primarily in causal terms governed by genes, with little attention paid to the intermediate roles of mind and culture. As a result, debates have often been polarized between those who stress the importance of consciousness and society and those who stress a more causal story in traditional scientific terms, even condoning forms of genetic determinism.

The cognitive science of religion, a new discipline which has developed very rapidly over the last few years, has inevitably been embroiled in this ongoing debate (Trigg and Barrett, 2014; White,2021). Such cognitive science tends to look for universals in human behaviour across time and culture, and to look for causal influences, perhaps to be explained through neo-Darwinian theories of evolution. Yet religion in all its diversity seems to be quintessentially a social activity with shared meanings and understandings that differ from society to society. Reducing it all to a common basis seems simple-minded. Yet the very fact we continue to talk of human sciences suggests that perhaps we do assume that all humans have something in common.

The later Wittgenstein's views have much to contribute to this debate, with various strengths and weaknesses fully exposed. Part of the key to his views lies in seeing what he was reacting against. In his *Tractatus Logico-Philosophicus* Wittgenstein (1933) had espoused what seemed to be a very scientistic view reducing all that could be said to what was within the reach of science. Before the Second World War, he was thought by members of the Vienna Circle, with their stress on a 'scientific world conception' to be a supporter of their views. For them, all metaphysical statements were meaningless, and, in particular, religious statements were nonsensical because they failed to engage with any reality[2]. The crucial feature of Wittgenstein's original views lay in the strong distinction he made between language and reality. For him, at that point, all language was to be understood on the model of name and thing named, a view explicitly repudiated in his later *Philosophical Investigations.*

While the early Wittgenstein saw language as essentially corresponding with parts of reality existing independently, the later Wittgenstein looked for the meaning that people gave language in the context of their practices and ways of life. This was particularly important in the case of religious statements, as many of his followers realized, because it meant that no religion could be criticized on the grounds that it was not about anything. Instead, it was important that it be understood how believers used their words and how they were linked to their actions in a wider context. Believers could thus never be 'wrong' or have false beliefs. They just lived a particular way of life. Wittgenstein at times used the term 'form of life' to describe the way in which concepts and actions were bound together in particular ways. For him, language could not be prised apart from any 'reality' but was just used in particular contexts. Meanings are not mental accompaniments to words, but we have to discover meanings by looking 'how the words in question *are actually used in our language*'.[3] The contexts are social because Wittgenstein stressed the public and shared nature of language. He famously denied that there could be a private language, one known only to an individual. We learn and teach languages in a public space through public activities.

This repudiation of a science-based interpretation of human activity in favour of a more human-centred appreciation of people's own beliefs and understandings ensured the priority of concepts over mere behaviour and events. When the social nature of concepts is stressed, it is perhaps obvious to link them with the fact of human language. When this is added to Wittgenstein's repudiation of any reality existing independently waiting to be named, language becomes not just the vehicle of thought but the means by which we construct reality. In *Zettel*,[4] fragments of his writings coming from the 1940s, Wittgenstein said that 'like everything metaphysical the harmony between thought and reality is to be found in the grammar of the language'. For Wittgenstein the notion of grammar is linked with the idea of a conceptual framework. He tells us[5] that 'grammar tells what kind of object anything is' and interestingly adds in parenthesis the thought that theology may be seen as grammar. Yet, again and again, Wittgenstein is concerned to repudiate the idea that we can have any private understandings or private meanings. Thought does not precede language but is constituted by language. He quotes[6] the report of William James in the nineteenth century about the recollections of 'a deaf-mute Mr Ballard' who allegedly had had thoughts about God and the world before he could learn to speak. This might be quite intelligible to anyone who imagines that thought and language are independent. For Wittgenstein however it created a problem, and in the end he just dismisses such recollections as 'a queer memory phenomenon' that does not entitle us to draw any conclusions.

The priority of language not only stresses the public, social and contextual nature of meaning. The mere fact that there are different languages begins to raise the question of what common ground if any human languages may have. Wittgenstein's own concentration on the concept of pain and how it is taught suggests that he is concerned with universal human phenomena, and indeed at one point he claims[7] to be supplying remarks on 'the natural history of human beings'. He does not seem to be talking about quirks of particular societies but instead is concentrating on fundamental facets of being human. The very fact that he talks of so-called 'primitive reactions' as forming the basis of various language-games, appears to appeal to a common human nature. Indeed he says in general terms[8] that 'the human body is the best picture of the human soul'.

1.2 Language-games and forms of life

While Wittgenstein does refer to general features of human life, his emphasis on language raises the question of the fact of different languages, different societies, different contexts and different ways of life. In fact, he says at one point:[9] 'I want to say: an education quite different from ours might also be the foundation for quite different concepts.' He goes on to remark that 'their life would run on differently' so that 'different concepts would no longer be unimaginable'.[10] This is an important point and raises the issue not just of different languages such as English or German. If our concepts are related to the ways we behave and different people behave differently with different ways of life, their various understandings may diverge markedly. The language of science may rely on different assumptions from that of religion so that the two languages become literally incommensurable. One can speak one and understand it and then use another form of language in a different context without necessarily putting them together, let alone translating the terms of one into those of the other. Wittgenstein himself was aware of this kind of possibility when he used the analogy[11] of seeing an ambiguous drawing of a 'duck-rabbit' first as a duck and then as a rabbit. It is difficult to see the drawing as both simultaneously. Yet he confusingly also explicitly wondered about the situation of an explorer coming upon a strange language in an unknown country. He says: 'The common behaviour of mankind is the system of reference by means of which we interpret an unknown language.'[12]

For Wittgenstein, concepts are integral parts of practices, and ideas are intimately linked with public ways of behaving, each with its own norms or rules. His use of the term 'language-game' illustrates the way in which he saw practices as a whole, so that a part could not be abstracted meaningfully from the whole. A kick of the ball in soccer or a batting stroke in cricket carries the meaning it does within the context of a particular kind of game with its own set of rules and purposes. Similarly, Wittgenstein uses the concept of the form of life to emphasize the role of collective agreements as the foundation of the practices that give language its meaning. When posing to himself the question of whether he is simply saying that human agreement decides what is true or false, he says that the agreement in the language used 'is not agreement in opinions but in form of life'.[13] When dealing with questions about what it is to be certain (in writings from the last year and a half of his life, 1949–51), he describes such certainty 'not as something akin to hastiness or superficiality, but as a form of life'.[14] Forms of life are a starting point and have to be accepted as given. Our concepts, as expressed in language, are linked to ways of acting. Their meaning is linked to the difference made to our lives.[15] As Wittgenstein says of a language-game, it is not based on rational grounds and cannot be reasonable or unreasonable. Instead he says:[16] 'It is there – like our life.' All we can say is that it is played.

As Wittgenstein himself had considered, all of this raises the question of truth. If agreement in judgements constituted truth, we would be on a slippery slope to an extreme relativism, given that disagreement is rife amongst human beings, even about very basic matters. On the other hand, if ways of acting are rooted in our nature as human beings, there will at least be some universality and commonality of meaning amongst humans, even in diverse cultures speaking different languages. That would provide a basis for translation and mutual understanding. In his final months, as Wittgenstein wrestled with these problems, it looks as if he became more inclined to accepting some form of conceptual relativism. In *On Certainty* for instance[17] he makes explicit the idea that different systems of belief can have their own internal standards so that there can be no non-question begging access to objective truth. Given his repudiation of the philosophical realism of the *Tractatus*, he can no longer think of language being about an independent reality, and therefore being misapplied. If it is linked to various forms of human behaviour, which can differ radically, their meaning and truth will vary according to language-game or form of life. This raises the question 'how can we distinguish between distinct language-games or recognize different forms of life?' Both terms suggest rule-governed activities based on integrated system of belief rooted in ways of acting. Both concepts raise the problem of diversity without resolving it. Is fundamental disagreement itself a criterion of such radical difference? If so, it would follow that apparent disagreement is not so much a contradiction of the other as evidence of a lack of comprehension.

The dilemma becomes very explicit in parts of Wittgenstein's *On Certainty*. In §106, he raises the question of what one would say if some adult told a child that he had been on the moon. He asks what reply he could make to the adults of a tribe who believe that people do sometimes go to the moon, perhaps interpreting their dreams in that way, but will indeed grant that there are no ordinary means of climbing up to it. He then asks the crunch question of whether there is no objective truth. Surely it is true or false that someone has been on the moon. His reply is that 'if we are thinking within our system, then it is a certain no one has ever been on the moon'. He furthermore rashly says that 'our whole system of physics forbids us to believe it'. This is in the context of his assertion (§105) that 'all testing, all confirmation and disconfirmation of a hypothesis takes place already with a system'. He continues: 'And this system is not a more or less arbitrary doubtful point of departure for all our arguments: no, it belongs to the essence of what we call an argument.' For him the system is the element in which arguments have their life.

Wittgenstein wishes to generalize this argument and in §107 asks: 'Isn't this all together like the way one can instruct a child to believe in a God or that none exists, and it will accordingly be able to produce apparently telling grounds for the one or the other?' Thus, religion is also treated as a separate system of belief with its own standards, without appeal to anything external to it to confirm or falsify it. For Wittgenstein we cannot have ungrounded presuppositions, merely ungrounded ways of acting (§110). It seems as if we will live a religiously based life and bring our children up in it or we will not. We will either behave as if people can get to the moon or we will not. Yet the moon example itself shows what a curious position that Wittgenstein is in. It seems very foolhardy to have said in 1950 that our whole system of physics forbids us to believe that people could get to the moon, or that physics forbids us from answering such questions as how the force of gravity can be overcome or how someone could live without an atmosphere (§108). It is all the more curious, as Wittgenstein's original academic studies before the First World War had been in aeronautical engineering. Whatever Wittgenstein's later thoughts about physics, within twenty years what he had claimed was clearly disproved and people did get to the moon. Reality and beliefs about it not only may not coincide but reality can have the last word.

1.3 Conceptual relativism

Wittgenstein also deals with the broader issue as to whether physics itself is reliable and a vehicle of truth. As already mentioned, he was reacting against the narrow views of the Vienna Circle which wanted to make physics the determinant of all truth.[18] The alternative danger, which it looks as if he could be falling into, is that rather than saying physics determines what is true, he retreats to the position that physics can no more claim objective truth than anything else.[19] He asks point blank whether it is wrong 'for me to be guided in my actions by the propositions of physics'. He thinks of those who do not think that physics gives good grounds and instead consult oracles. In a parenthesis, he says darkly that 'for that we consider them primitive'. Then he faces the basic question: 'Is it wrong for them to consult an oracle and be guided by it?'. His answer is that to call them wrong is simply to use 'our language game as a base from which to combat theirs'. He pursues the point by suggesting that all we can do is use slogans and call each other names. For instance,[20] he comments that when two principles really do meet that cannot be reconciled with one another, 'each man declares the other a fool and a heretic'. In the end, his position is that what counts as justified or unjustified depends on whether a language-game exists in which an assurance is offered.[21] What people consider reasonable or unreasonable can alter. The only question we can ask[22] is 'What difference does it make to their lives?' As Wittgenstein muses, it looks as if 'there is no objective character here'.[23] In the context of religion, in his *Lectures on Religious Belief*, Wittgenstein remarked that 'whether a thing is a blunder or not – it is a blunder in a particular system. Just as something is a blunder in a particular game and not in another'.[24] What counts as reasonable therefore depends entirely on its social setting.

The picture given in all this is that words are rooted in ways of acting, and systems of belief cannot be torn apart but must be accepted as a whole. There is nothing beyond them to which reason can appeal. What counts as reality is internal to our language-game or form of life. In all this the idea of system, language-game or form of life all seem to have similar roles but it is unclear how any particular one is to be identified. In the context of religion, disagreement in judgements may seem to make for a different basis of acting and therefore for a different form of life. Yet does that mean that religion as such is not a form of life but a particular religion might be? A branch of Christianity, such as Roman Catholicism might be seen to vary in its judgements sufficiently to be distinguished from Protestantism. Would they create different forms of life? So, one could go on, with enduring disagreements splitting up social groups further and further. We seem far removed from Wittgenstein's reference to the 'common behaviour of mankind'. Similarly, in the field of science, is physics to be for one form of life and chemistry another? Does a scientific outlook as such constitute a form of life compared with a religious one? If fundamental disagreement is always evidence of different forms of life, we all run the risk of finding ourselves in a form of life comprising one person. That is emphatically what Wittgenstein is denying with his emphasis on the public and social nature of concepts.

This idea of different conceptual systems, whether physics and oracles, science and religion, or anything else, inevitably becomes a type of conceptual relativism that regards such systems as insulated from each other. There could be no common external standard, understood by all, that could be appealed to. The result is that they will be incommensurable with each other in much the same way that Thomas Kuhn viewed different scientific theories before and after scientific revolutions[25]. One is either a believer and understands, or an unbeliever and finds the opponent's beliefs incomprehensible.

This concentration on human diversity and pluralism, making apparent fundamental disagreement into evidence of different conceptual schemes, undermines any possibility of rational comparison of different views. The idea of some dispassionate reason, that is able to detach itself from context, and attempt impartial judgement, is totally ruled out. The later Wittgenstein is certainly on the side of those who maintain that a so-called God's eye view is in principle impossible. It is not surprising that his views have been one of the formative influences on the growth of the postmodernist rejection of modernist ideas of rationality. Indeed, in *On Certainty* (§475) he explicitly says that he wants 'to regard man here as an animal; as a primitive being to which one grants instinct but not ratiocination (*Raisonnnement* in the original German)'. He adds emphatically that 'language did not emerge from some kind of ratiocination'. While it may be significant that he still wants talk of 'man' (humanity) in general, this refusal to see humans as 'rational animals', or reason as existing independently of different languages, entails a significant downgrading of the role of reason in human life, with dire implications for any idea of philosophy, let alone metaphysics. For him, reasoning did not undergird language but emerges from it. Since languages are of their nature diverse, this immediately splinters human rationality into myriad pieces.

The rejection of any commonality lying outside and beyond differing societies and languages raises questions about the coherence of Wittgenstein's stress on instincts and the natural history of human beings with the extreme diversity he appears to encourage. As post-modernism has come to recognize, the very notion of a common humanity can be challenged. Particular identities are more and more stressed. Once diversity and disagreement become the ultimate facts of human existence, we cannot transcend them but become seemingly trapped within whatever identity some form of life has presented us with. As Wittgenstein himself was at pains to argue, the idea of a reality existing beyond language and serving as its focus and basis become suspect and cannot be relied on as a common goal.

The twin buttresses of a common human nature and objective reality that confront us all could offer support for a robust understanding of human reason, able to transcend its particular circumstances. The whole purpose of Wittgenstein's later philosophy, and his concentration on language and its uses, was to repudiate appeals to an objective reality. He quite rightly opposed a narrow scientism, as championed by the Vienna Circle. Yet the consequence of the concentration on alternative uses of language was to deny the possibility of even science itself being able to claim truth. His suspicion of physicists' claims that they could describe objective reality says it all. In the end, Wittgenstein argued for the legitimacy of many different forms of language by debasing the significance of all of them. None of them could claim truth. Science came to be one arbitrary system of belief amongst many alternative human practices.

Wittgenstein's regular references to religion and theology, sometimes in parenthesis, shows that he was willing to make room for religious statements and constantly had them in mind. Yet it was at the cost of removing any possibility that they too could claim truth. Whether in science or religion, if one cannot believe that what one is saying is true, the question arises whether there is any point in believing or saying it. Just because it is a constituent part of my practice or our form of life does not seem sufficient justification. Sincerely held religious belief, when put into practice, can often be costly in many ways. It can even invite persecution. Why continue to hold it, unless truth appears to be at stake? Any position that undermines faith in human reason can soon arrive at a nihilist conclusion. Why, too, practice science at all, if it is just one way of living a life amongst many? The narrowness of identifying human rationality with the practice of modern science may be legitimately questioned, but not at the expense of denying the very idea of rationality.

1.4 Wittgenstein and cognitive science

What is the relevance, if any, of Wittgenstein's later thought to the cognitive science of religion? Like all science, it gains its point from assuming one real world that is the same everywhere. Science thrives on the belief that what is true in Beijing will be true in Washington DC. What is true in London will be true in Nairobi. Yet if science is relegated to be just one distinctively Western practice, it loses its point. As well as assuming the universality and objectivity of one reality, one world common to all, cognitive science of religion must also presuppose that its researchers can find commonalities between different human beings in many different cultures. If it only gives us facts about modern Western societies, it is worthless. The concept of universal human nature enables it to generalize from one culture to another, and from one time to another. It enables anthropologists to investigate one society and draw conclusions about humanity everywhere. It enables them to predict from findings in Western society what could be the truth in China, with a radically different religious history, and empirical findings can bear that out.

All human cultures answer to the same basic and even human needs and interests. Humans everywhere need to eat, drink and have shelter, but, in addition, cognitive science draws attention to the fact that the workings of the human mind ensure that humans are not pinned down by instinct or basic biological needs.[26] The human mind typically can range widely. The very existence of the cognitive science of religion as a rational enterprise bears witness to this. In the same spirit it can point out that there are common cognitive patterns rising in very different contexts. Across cultures humans consider the same questions. Much modern cognitive science of religion has been devoted to showing how, for instance, humans typically look for purpose in events. They find it easier to believe that things happen for a reason than that they are haphazardly caused. Children will say that rocks are pointed so that birds cannot sit on them and find that easier an explanation to embrace than sophisticated ones about sedimentation and erosion. Similarly, humans typically look for a non-purposive agency when they cannot see physical agents. This is no doubt the origin of beliefs in tree spirits or in a god of the sea who controls its power. People across cultures find it difficult to envisage death as being the end of everything and find it natural to think of their ancestors surviving. As has often been pointed out, we are by nature dualists able to envisage the separation of mind and body.

All these claims, which are the product of years of careful empirical research, are assumed to have cross-cultural validity. The cognitive science of religion is opposed to any notion of social construction whereby the concepts of our society determine for us what should be real. The philosophical assumption behind such cognitive science is that how we think does not determine how we see the world. It is rather the result of the effect of the world and its problems on us, and a human understanding which has perhaps developed through evolution. The kind of examples just given do not themselves constitute a religious outlook but certainly provide the building blocks for such a view. They incline us to look beyond the world of appearance and search for deeper explanations. That is the motivation of contemporary science, but it all too often sets itself limitations, based on implicitly philosophical views about the kind of things that can exist. That is what the Vienna Circle did. Science of its nature may only be concerned with physical interactions, but it should not be tempted into assuming that physical reality is necessarily the only kind of reality. The cognitive science of religion certainly suggests that humans across time and space have not believed that.

One way of putting all of this is that the cognitive science of religion has discovered a God-shaped hole in human conceptual understanding. It can be filled by many kinds of diverse views about God or gods. The many religions of this world fit into that space.[27] While we are ready and primed to look for non-physical agency and evidence of purpose in the process of the world, such cognitive science shows what human beings find it easy to understand and believe. It shows what they are looking for, and what questions they want answered. It does not prescribe what should be found or indeed whether there is anything to be found.

Issues concerning child development are relevant to cognitive science, and again research will look at children in many different cultures expecting to find commonalities. For example, cognitive science reports that, as children develop, they find it easy to embrace the idea of a God who knows everything. One simple experiment with small children indicates this. At the age of three, children can be shown two cups. An apple is placed under one of them when the mother is watching, and then she leaves the room. The children assume that the mother knows under which cup the apple has been placed. If then the apple is placed instead under the other cup, when she is still out of the room, children of that age will naturally say that she knows where the apple now is. Mummy knows everything. Intriguingly, about a year later, at the age of four, as their minds have developed, children acquire what psychologists call a 'theory of mind', according to which they realize that other people's perspectives may not be the same as our own. They therefore realize that if the mother did not herself see where the apple is now placed, she will now wrongly still believe it is under the original cup. Yet if such children are asked whether God would know where the apple now is, they will be quite sure that He would. The point of this experiment is not about children's upbringing or what they are taught about God. It is that children find it easy to think of God knowing everything even when they come to recognize that ordinary people, even their mother, do not. As has been pointed out by cognitive scientists of religion, this continuing idea of an all-knowing God can itself have a salutary effect on human behaviour, just as the knowledge that we are being watched constrains what we do on occasion.

The cognitive science of religion is never about what could be true or false. It is about what humans find it easy to believe and the character of typical human tendencies to beliefs of a certain kind. Such cognitive science makes substantive claims about human nature. It cannot make further claims about the truth or falsity of beliefs without bringing in particular philosophical assumptions, perhaps itself buying into philosophical naturalism. What cognitive science does show is that the basis of religious belief lies deep in human nature and that it fits in with basic tendencies of human thought. While it grossly under-determines which religious beliefs are adopted, the cognitive science of religion can point out the natural building blocks of a religious attitude to life. It suggests why we might be primed, as human beings, to accept revelations of the nature of divinity. It shows why some religion, of whatever kind, is likely to be prevalent in all human societies and will be difficult to suppress in the long term, even by vicious totalitarianism. Basic religious responses occur at an instinctive level in human beings. They are 'first-order' responses whose significance and meaning have to be elaborated.

Rational reflection on human circumstances is certainly a feature of human life but is a second-order activity. What is being claimed is that religion, however defined, is what might be called a primitive impulse in humans. This fits in with what Wittgenstein might have been thinking about when he referred to 'the natural history' of human beings and their 'primitive reactions'. Elaboration on the significance of all this is characteristically a second-order activity, engaging human rationality and reflection. That is the role of theology. In the same way, modern science is a second-order activity reflecting on our basic experience of the world as human beings. Science and theology, are both complicated enterprises, growing out of teaching and learning. They thrive on the assumption that they are each in the business of reflecting on what is true. Basic religious impulses are there as part of our life.

1.5 Cognitive science and reductionism

One problem with the cognitive science of religion is the way in which some researchers are quick to link cognitive structures to so-called modules of the brain, or other neural structures, identifying them with physical processes produced by evolution. Empirical findings of such cognitive science then become overlaid with philosophical doctrines about the relationship of the mind and the brain, and the possible reduction of concepts to states of the brain. The discovery of universal traits of human cognitive thought should be kept distinct from questions about their origin and development. Whether conceptual thinking is hardwired into neural processes and can be explained by neo-Darwinian doctrines of genetic transmission is an important issue. Yet such views do not challenge the idea of universal human nature and take for granted the idea of an independent world to be investigated.

The fundamental problem remains whether we can steer a middle course between ideas of social construction, which imagine that concepts are wholly rooted in forms of social life, and a physical reductionism that turns the investigation of the nature of human thought over to the realm of neuroscience, evolutionary biology or a combination of the two. Such a scientistic approach operates within a philosophical naturalism that undervalues the role of human concepts in human life by suggesting that concepts do not operate independently of physical processes but are determined by them. Paradoxically such an approach is itself highly conceptual, and in claiming truth runs the risk of undercutting itself. Is a belief in the rootedness of conceptual thought in brain activity itself merely an epiphenomenon of brain activity? How can we conceive of what is true, as distinct from merely being conditioned one way or another into holding such beliefs? The reduction of conceptual thought to something else is fraught with difficulty and the threat of self-contradiction.

The approach of the later Wittgenstein and of some tendencies in contemporary cognitive science of religion each carry their own, contrasting, dangers. While science seems firmly grounded in its belief that there is such a thing as human nature and the common reality we all face, it seems to neglect the free-ranging nature of human reason and the independence of the conceptual thinking on which it depends. On the other hand, Wittgenstein is right to stress conceptual understanding and the independence of concepts, but by tying them too closely to particular societies and emphasizing their diversity he runs the risk of a debilitating relativism. Particularly in *On Certainty*, he seems to forget the crucial importance of anchoring concepts in something beyond their social surroundings. If all concepts are socially constructed, and there is no common frame of reference in which they can be compared, we all become imprisoned in our own ways of thought. Any possibility of dialogue between those who disagree becomes impossible. Indeed, the very concept of disagreement at a fundamental level is dissolved. People may have divergent concepts, but they will no longer be able to contradict each other or argue rationally. Reason cannot then transcend language – games or forms of life, as there will be no conceptual resources with which to do so.

Some forms of cognitive science try to explain away widespread ideas of what is true while still paradoxically proclaiming their own ideas of where the truth lies. Wittgenstein on the other hand challenged the narrow scientistic ideas of the Vienna Circle with which he was once associated. However, he tends to take all human conceptual schemes at face value, in accordance with our self-understanding, and refuses to accept that they too are subject to rational examination. He upholds the importance of human conceptual understanding in the face of scientistic attacks and resists narrow scientific views of reason. At one place[28] he says explicitly that 'no supposition seems to be more natural than that there is no process in the brain correlated with . . . thinking'. His conclusion is that it is possible that 'certain psychological phenomena *cannot* be investigated physiologically, because physiologically nothing corresponds to them', Wittgenstein thus recognizes both the independence of conceptual thought and its priority. Yet his lack of grip on the idea that concepts are trying to reflect something independent of them means that conceptual thought takes on a priority that respects diversity, without the ability to appeal to some overarching rationality that can still stand in judgement on it.

Wittgenstein's later thought takes us in a direction of disintegration into different conceptual schemes that can be further split into myriad pieces, given enough disagreements. Yet the tension between the thought of the later Wittgenstein and modern theories of cognitive science illustrates a problem. If all conceptual thought is always regarded as valid in its own terms, that in effect is the end of science. Even physics becomes merely the conceptual scheme of a particular kind of society. Yet on the other hand, science can overreach itself, when cognitive science, particularly the cognitive science of religion, tries to explain away basic conceptual schemes. The effect is to undermine their ability to claim truth, even though cognitive science assumes that it itself still can. It can enlighten us about human tendencies and dispositions and demonstrate the universality of impulses underlying religion. Yet it can do nothing to explain the elaboration of such impulses into doctrines and grand theological schemes. Similarly, it cannot explain its own ability to reach out to what it regards as true. Wittgenstein warns us of the dangers of a narrow naturalism that does not take human conceptual thought seriously as an independent force. The cognitive science of religion reminds us that such conceptual thought should not just be taken at face value but can have its roots in basic features of human nature.

We lose the notion of objective truth at our peril, even if we have to be painfully aware that none of us are in full possession of it whether in science or elsewhere. Diversity of thought is always a challenge, but not something meekly to be accepted at face value without proper rational consideration. Reason matters, and for that we need the conceptual thinking with which Wittgenstein was preoccupied. We have to remember that such thought always purports to be about something. When reacting against hubristic claims about science and scientific progress we must not lapse into an opposite extreme that flirts with scepticism about the possibility of any overarching rationality or truth. Such a view in the end destroys both science and philosophy.

Notes

1 Trigg 1982, 2001.
2 Trigg 2015.
3 1958, 56.
4 1967, §55.
5 1963, §373.
6 1963, §342.
7 1963, §415.
8 1963, iv.
9 1967, §387.
10 1967, §388.
11 1963, xi, 194.
12 1963, §206.
13 1963, §241.
14 1969, §358.
15 1969,, §338.
16 1969,, §559.
17 §105.
18 See Trigg 2015, 6 for a discussion of the Vienna Circle's views.
19 1969, §608.
20 1969, §611.
21 1969, §620.
22 1969, §338.
23 1969, §336.
24 1966, 59.
25 Trigg 1973.
26 White 2021.
27 Trigg 2020.
28 *Zettel*, §608.

2

On truth, language and objectivity

Florian Franken Figueiredo

2.1 Introduction

During the 1950s and 1960s a scientific movement, often called the 'cognitive revolution', arose. Leading voices in this movement claimed that research into the mind and the brain should focus on the study of processes in the brain and the vast majority of these philosophers and scientists took their guiding model of the mind and brain to be computational and representational.[1] Their intellectual efforts resulted in the establishment of the cognitive sciences and, in the 1990s, in the establishment of the cognitive science of religion (CSR) as a subdiscipline.[2]

The early stages of CSR were accompanied by its proponents' dissatisfaction with contemporary trends in cultural studies – a development that will not be the focus of this chapter. Suffice to say that proponents of CSR are mostly united in challenging the following presumptions, commitments and accounts that they believe provide the basis for cultural studies in the 1990s:[3] (1) the postmodernist claim that the world is the result of multiple narratives and, as such, is a social construct, (2) the cultural relativist assumption that a person's behaviour is only relative to their particular culture and thus can only be understood in light of that specific culture, (3) the cultural determinist belief that human behaviour is more or less determined by its sociocultural environment, (4) the interpretative account of religious ideas and behaviours that views culture as a semiotic system of signs and symbols while suggesting that the interpretation of human actions should take the place of explanatory approaches, (5) accounts that exclude the mind in seeking to explain human thought and behaviour (such as Freudian psychodynamic theories and the 'black box' theory of human psychology), (6) accounts that privilege religious diversity and differences between religions over cross-cultural patterns without attempting to situate them in an explanatory framework and (7) accounts that are resistant to those religious studies that advocate reductionism because they assume that religion cannot be reduced or compared to anything other than itself.

Before I relate these preliminary remarks about CSR to Roger Trigg's both illuminating and thought-provoking chapter of this book about Wittgenstein, concepts and human nature, it should be pointed out that while proponents of CSR challenge the presumptions, commitments and accounts that I have just mentioned they mostly agree on a set of basic assumptions that form the methodological framework of their own studies. Most important for this framework is the assumption that (1) all human behaviour can and should be theoretically explained. Regarding the nature of theoretical explanations their proponents furthermore agree that the explanations pursued should (2) be objective, (3) focus on the processes in the mind/brain as being the causal factors of human behaviour, (4) be open to empirical assessment. Such a framework entails (5) a reductionist account in which causation is traced back from higher to lower levels (e.g. the behaviour of gases can be explained in terms of motions of molecules).

It will not come as a surprise that Trigg, as he is also a proponent of CSR, shares these assumptions, or at least most of them. What is excellent about his chapter, however, is that he does not take them for granted but that he critically reflects on their limitations. He seeks to raise awareness of the possibility that by following the assumptions blindly one might find oneself in danger of pursuing what he calls a 'narrow naturalism'. Trigg accepts that religion is a social activity with shared meanings and understandings and he points out that studies in CSR should not undervalue the role of human concepts in human life. On the one hand he thinks that Wittgenstein's later work, which he characterizes as being oriented to human beings and their activities, has to a certain extent the potential to highlight the importance of socially shared concepts for our understanding of religion. On the other hand he sees great difficulties with Wittgenstein's approach as he holds that it shares common ground with the aforementioned presumptions, commitments and accounts in cultural studies that are usually rejected by proponents of CSR.

I have much sympathy for Trigg's view that an adequate understanding of naturalism implies a strong distinction between the origin and development of biological processes that are the natural basis for human thought and the question as to whether the contents of human thought reveal universal traits. He argues that the later Wittgenstein should not be conceived of as a helpful advocate in arguing for this distinction as he interprets Wittgenstein's ideas as supporting some sort of postmodernist account. In this chapter I present a reading of Wittgenstein's ideas that differs in important respects from the one put forward by Trigg. I specifically argue that my alternative understanding precisely suggests the distinction he aims to make but avoids misunderstanding and thereby undervaluing the role that the physical world plays in our understanding of human behaviour. In other words, I argue in this chapter that Wittgenstein's ideas are rather helpful in finding that very middle course between social construction and physical reductionism that Trigg is looking for.

2.2 The contrast between two Wittgensteins – early and late

Until the 1980s, the common understanding often found in introductions to, or encyclopaedia articles about, Wittgenstein's philosophy was that there is an early period of Wittgenstein's work summarized in the *Tractatus Logico-Philosophicus* (published in 1921) followed by a later period that was taken to be mainly represented in the work of the *Philosophical Investigations* (published posthumously in 1953). Both periods were often perceived as two distinct stages of Wittgenstein's thought. One factor that supported this view is the interpretation of Wittgenstein's own remark in his preface to the *Philosophical Investigations* in which he suggests that these new ideas 'could be seen in the right light only by contrast with and against the background of my older way of thinking [i.e., the one presented in the *Tractatus*]'.[4]

Trigg follows this standard account suggesting that a reductionist ('scientistic') view is to be identified with the earlier stage of Wittgenstein's work whereas a conception of the world as a social construction is to be found in the later stage. He interprets the contrast between these two stages of Wittgenstein's ideas as a matter of the later stage repudiating the earlier reductionist view at the costs of (a) giving up the general idea that language pictures an independently existing reality and (b) adopting the idea that reality is constructed by the use of language in social contexts. Even though Wittgenstein speaks about the contrast between the ideas in the earlier and the later book, it is however quite misleading to understand this as demarcating two different stages in Wittgenstein's work, as these two stages do not actually exist. This is one important point at which Trigg's analogy fails. Scholars of Wittgenstein have increasingly argued that his work is best understood as continuously evolving throughout his life – as evidenced in the development of his thought in the *Nachlass* – rather than as a matter of two contrasting stages.[5] Wittgenstein's remark in his preface is thereby misunderstood when used as a guide to the clarification of his philosophical development. Rather, it should be taken as a piece of heuristic advice about how to better understand his new ideas which are the product of many single steps in his previous thinking.

That said, there is textual evidence for the claims that, between 1929 and 1938, Wittgenstein gives up the Tractarian picture theory and that his later philosophy is guided by a different set of analogies than the one he puts forward in the *Tractatus*. According to the picture theory, the proposition is a picture of reality and, as such, 'is laid against reality like a measure'.[6] That this view is constantly changing becomes clear from a remark that Wittgenstein makes to members of the Vienna Circle in December 1929. He explains to them that he now thinks that it is 'an entire system of propositions that is compared with reality, not a single proposition'.[7] Later in his development Wittgenstein becomes aware that, in contrast to what he has claimed in the *Tractatus*,[8] the rules for projecting propositions onto reality are not implicit in the representing relation. In 1938 Wittgenstein writes that what underlies the misguided idea is to think of the representing relation as projection lines and to confuse them with the method of projection.[9]

It is controversial amongst Wittgenstein scholars whether the author of the *Tractatus* intends to make, or is committed to, any specific ontological statement about reality. Even if the answer to this question is affirmative, it seems safe to say that the picture theory does not pursue the goal of making such an ontological statement as its focus is not on questions concerning reality per se but on a supposed picturing relation between language and reality. The picture theory is concerned with logic as a theory of the relation between language and reality and thus its rejection should be conceived of as independent from ontological questions about reality. Wittgenstein proposes an account of the nature of meaning based on the presupposition of reality (i.e. based on the existence and non-existence of states of affairs: a presupposition that is never questioned). In early 1930 one can find Wittgenstein's first attempts to formulate a different account that provides an answer to the same question but from a different perspective. The account is based on the idea that the 'meaning of a sentence is its purpose'.[10] Again, here the focus is not on reality per se but rather on the use of words and sentences and what we use these for. Wittgenstein keeps this idea in the *Philosophical Investigations*, neither denying the existence of a reality nor claiming that reality is constructed by the use of words and sentences.

It is important to appreciate that any claim that the use of language constructs reality is based on a commitment to the assumption that language and reality are two separate realms or entities and that there is a philosophical theory – possibly not dissimilar to the Tractarian picture theory – that may tell us how they are related to one another. For the later Wittgenstein this is a misguided idea. According to him, no theory is needed in philosophy that is supposed to explain the agreement of either language and reality or language and action. The temptation for such a theory is motivated by the confused idea that any suggested 'agreement' requires a theoretical explanation. Much of Wittgenstein's philosophy is about arguing that there is no gap between language and reality that needs to be filled by either a theory of projection or a theory of use. The use of language is always use *in reality*.

Wittgenstein's example of this point at the beginning of the *Philosophical Investigations* is a situation in which children are brought up 'to perform *these* actions, to use *these* words [. . .] and to react in *this* way to the words of others'.[11] It is important here to notice the ostensive use in this situation (indicated by the italics in the original). The ostensive gesture, as one might be inclined to think, does not establish a relation between language and reality but it is part of reality. A child does not construct reality but it is in reality that it learns to talk. Here the point is that reality is not a product of explanation but the context in which children learn to perform actions and use words. Any 'ostensive explanation' or 'ostensive definition', as Wittgenstein calls it, would not be successful if children had not already been trained in the context of reality how to use the pointing gesture.[12]

In sum, for the later Wittgenstein reality is *not* a construct. Wittgenstein's term 'language-game' is misunderstood when one thinks that it constructs reality. The term rather informs our understanding of what we mean by the concept of reality, exploiting the fact that language-games *are* part of reality, unexplained, 'there – like our life'.[13]

2.3 The social nature of meaning

Wittgenstein's opening remarks in the *Philosophical Investigations* demonstrate that a basic language activity such as naming is philosophically misunderstood in the framework of the idea that language is projected onto reality. What can be observed instead is that naming is one of the ways that words are used, for example, when children are learning language. Augustine was not wrong in his assumption that he learnt from his parents how to name objects. He was wrong *inter alia* in assuming that what underlies the learning process is the specific relation between names and named objects and that he learnt language by observing how this relation was manifested in his parents' utterances.[14] Wittgenstein argues that things we do with language, such as naming, are social activities. Words can be used to name objects, as is illustrated by his famous example of the builders: a builder calls out the words 'block', 'pillar', 'slab' and 'beam' and his assistant brings the stone which he has learnt to bring at such-and-such a call.[15] What happens in these kinds of scenarios is that *an activity* is carried out or engaged in. As Wittgenstein describes it, naming 'is rather like attaching a name tag to a thing';[16] that is, it is an interaction between two or more people in which one person uses a word by naming an object and the rest either has to decide or is instructed in a particular way of how to go on.

In the Tractarian picture theory the meaning of names is explained by the existence of objects. The idea is that names have meaning because they picture reality and that the rules of language application are implicit in the picturing relation. According to the later Wittgenstein, to hold that the normative source of language is some supposed picturing relation between word and world is to misunderstand what it is to be a language user. Might there be an alternative explanation though of what guides our meaningful language activities? If naming is characterized as a social activity like attaching a name tag, are there any persistent rules involved in attaching those name tags? If we accept Wittgenstein's view that being a language user relies on instruction and training then, it seems, we also have to accept that different instructions lead to different understandings of particular concepts. Does this not lead us to different understandings that, at some point, are *incommensurable* with another? How can we explain that people understand each other, at all?

What motivates these questions might be a scenario like the following. A teacher gives one of his pupils the order to write down series of signs according to a certain formation rule. Let us assume that, judged by the usual criteria, the pupil has mastered the series of natural numbers. Now the pupil should write down series of the form '0, n, 2n, 3n, . . . ' at an order of the form '+n'.[17] Let us further suppose that the teacher has done exercises with him and tested his understanding up to 1000. Subsequently, the teacher gets the pupil to continue the series '+2' beyond 1000 and the pupil writes '1000, 1004, 1008, 1012'. When he is asked by the teacher to check what he is doing, he does not understand. Perhaps the teacher says: 'You should have added *two*: look how you began the series!' And the pupil answers: 'Yes, isn't it right? I thought that was how I *had* to do it.' He might even point at the series and insist: 'But I did go on in the same way.'

It seems safe to say that people will want to agree with the teacher and say that what the pupil does is wrong. Perhaps one wants to say that the pupil does not follow the rule properly or that he does not understand the order correctly. The idea of letting the pupil get away with it and accepting that how he reacts to the teacher's order is a somehow accurate understanding of the order not only seems ludicrous but also raises the reasonable question 'why should it count as an accurate understanding if it is incommensurable with ours?'[18] In order to explain the meaning of expressions like 'add two', or the meaning of *signs* in general (including mathematical signs like '+2'), one may think that some other guiding relation is being used, that is somehow independent of the performed action. I have mentioned that later Wittgenstein has become aware that this idea heads in the wrong direction. One might object, however, that social practices seem unable to set the standards for correct understanding either. For not only is it possible that different instructions produce different ways of understanding what is meant but, as the example demonstrates, it is also the case the instructions do not guarantee that they are followed correctly. This is pointing to the fact that it is unclear what it means to follow an order correctly in cases like this one.[19]

So why should we nevertheless assume that the normative roots for our understanding lie in our social practices? First, there is observable evidence for the fact that the way in which young human beings acquire certain understanding is by getting instructions. More importantly, particular instructions establish the very meaning of what people do in these practices. As Wittgenstein puts it: 'We got to know the *nature* of calculating by learning to calculate. [. . .] *This* is how one calculates. Calculating is *this*. What we learn at school, for example.'[20] Practices often lack meaning for someone who has not been introduced into them by means of instruction. This, as Trigg rightly points out, is a valuable lesson especially for anthropologists who try studying a society from the security of their armchair.

Second, if one follows the idea, put forward by Wittgenstein, that words and sentences have meaning in so far as they are used for particular purposes under certain circumstances (which is, importantly, not the same as to say that meaning *is* use), then social practices provide the adequate framework for the proper understanding of the normative roots of language. The sign '+2' can be used for different purposes just as handles in the cabin of a locomotive do different things and can be used in different ways:

> one is the handle of a crank, which can be moved continuously (it regulates the opening of a valve); another is the handle of a switch, which has only two operative positions: it is either off or on; a third is the handle of a brake-lever, the harder one pulls on it, the harder the braking; a fourth, the handle of a pump: it has an effect only so long as it is moved to and fro.[21]

Without certain instructions it is not even necessarily the case that '+2' must be understood as either a sign that is used, in a strict sense, to do mathematics or that the sign's meaning is similar to the one of the order 'add two'. The sign can be used, for example, in a supermarket as an invitation to take two items instead of one. Whether one can call it a name or a sentence is also difficult to say – at least at the face of it.[22] How the sign is used, as the example demonstrates, rather depends on the circumstances and these are often similar to the circumstances in which we have been instructed to use the sign in certain ways.[23] In the example, the pupil is instructed to engage in the '+2' activity up to 1000. Beyond 1000, however, there is no agreement amongst the agents, pupil and teacher, as to how the sign is supposed to be used in these cases. No instruction took place that would have established such an agreement. The teacher assumes blindly that the pupil will follow *his* use of the sign in these cases. The pupil, however, surprises the teacher by putting forward a different suggestion in writing '1000, 1004, 1008, 1012'. Now the agents have to somehow come to an agreement as to how the sign should be used for the given purpose and under the given circumstances (imagine that the role between teacher and pupil changes and the pupil is going to give instructions to the teacher of how he, the pupil, uses the sign in these cases).

But is the teacher not right in assuming that the pupil will follow his use of the sign in continuing the series beyond 1000? Is this not the way in which we usually do it? These questions refer back to the initial reaction that perhaps most people will have when they are confronted with the scenario. The more profound philosophical questions, however, are about the justification of what is assumed here. What justifies the pupil in using '+2' in a particular way under the given circumstances? In the *Philosophical Investigations*, Wittgenstein explores various possible answers to this question at length. One might think, for example, that the use is justified by the agent's inner process or inner state, for example, by an intuition, a state of consciousness, a mental state, the neural configuration or neural activity of one's brain and so on.[24] Wittgenstein argues instead that language-involving activities are justified within social practices in virtue of an agreement between the interacting agents.[25]

This does not mean that agents always find themselves in a situation where they need first to come to an agreement of how to use language and how to go on correctly under certain circumstances. According to Wittgenstein '[i]t is not only agreement in definitions, but also (odd as it may sound) agreement in judgements that is required for communication by means of language.'[26] The fact that in many ways we already agree in our judgements and how to form them might explain the reaction of most people whose first impression is that they want to agree with the teacher's use. Take the signpost as another example of our agreement in our judgement and practices. People usually do not negotiate an agreement in front of a signpost as to how to use it. We have learnt by instruction what it means to point at something and how the signpost is used for that purpose (i.e. how it indicates the direction[27]). We have learnt how to follow the signpost in the correct way so that a different use of it would seem strange to us. However, we might imagine that, under certain circumstances, agents come to an agreement for a new use for the signpost and give new instructions on how to use it in a different way than we are used to.

2.4 Is there a problem with relativism, and what is it?

Following the argument so far, one might be inclined to assume that it leads to a form of conceptual relativism. If it is up to the agents to negotiate agreements about how to use language then the meaning of, for example, words and sentences seem to be entirely dependent on our agreements. This appears to be in tension with our assumption that the meaning of words for physical objects depends on the existence of those objects, that is on reality. As I have argued, Wittgenstein does not deny the fact that language-involving activities occur in reality and that the circumstances in which expressions are used play an important role in how their meaning is understood. He argues that the use is not determined by anything other than the framework of socially shared practices in which agents suggest, negotiate and instruct each other how to act and how to go on under certain circumstances.

Does this mean that Wittgenstein accepts a form of conceptual relativism? The answer surely depends on how the term is understood. The standard interpretation of relativism is explained in terms of truth.[28] Here is how Trigg relates the interpretation to Wittgenstein:

> [D]ifferent systems of belief can have their own internal standards, so that there can be no non-question begging access to objective truth. Given [Wittgenstein's] repudiation of the philosophical realism of the *Tractatus*, he can no longer think of language being about an independent reality, and therefore being misapplied. If it is linked to various forms of human behaviour, which can differ radically, their meaning and truth will vary according to language-game or form of life.[29]

If the notion of relativism means that the meaning of concepts is relative to their use for particular purposes under certain circumstances then Wittgenstein might agree, although he would not take this as a theoretical insight but as a proper description of how we use language. It is difficult to see, however, how Wittgenstein's account leaves room for forms of extreme relativism, incommensurability and the impossibility of mutual understanding, let alone for the claim that his position implies or even entails them. Again, Wittgenstein emphasizes the point that it is 'agreement in judgements that is required for communication by means of language.'[30] Sometimes, but by no means always, the agreement requires that negotiation is required amongst interacting peers, and sometimes negotiations might not lead to mutual agreement.[31] Yet there is no place for extreme relativism in Wittgenstein's account. For one, the social nature of his account implies the possibility that the occurrence of new instructions and negotiations might lead to understanding, eventually. There is no reason to suppose that understanding is *conceptually* excluded between human interacting peers who are assumed, at least, to be capable of understanding each other. For another, as Trigg would admit, the assumption of disagreement presupposes that there is always already some sort of agreement amongst peers from where negotiations may start. Those agreements may, but do not need extend to the level of commonly shared natural (or 'primitive') reactions which might be conceived of as belonging to 'the natural history of human beings.'[32]

If it is not the social nature of meaning that informs concerns about extreme relativism then what might it be? I have argued that later Wittgenstein rejects the idea that the relation between language and reality has explanatory (and within a theory of use: justificatory[33]) force that serves a theory of meaning. He changes from the approach of trying to give an explanatory theory of how things really are (i.e. independent from us as human beings) to the seemingly less ambitious, but actually more philosophically insightful approach of describing the (human) understanding of how things are for us (as human beings) given the circumstances in which we (human beings) are. This means giving up on the idea that meaning is to be explained by independent factors and, in particular, an independent reality. What might seem to be the weaker approach is one that not only makes no attempt to determine independent factors but offers the description of a framework in which meaning is established. This approach might give the (mistaken) impression that the ways in which meaning is established in this framework are completely arbitrary and detached from existing reality. Two important things, however, are being overlooked: First, the ways in which meaning is established are manifested in a non-arbitrary structure that Wittgenstein calls 'grammar'.[34] Second, identifying Wittgenstein's account with some kind of conceptual relativism that is detached from reality implies a commitment to the idea that an independent reality is somehow related to language. I have argued that this is an illusion. Philosophers who feel a strong commitment to the assumption of a language-independent reality will have stronger difficulties to see through it.[35]

2.5 The concept of a system

The Tractarian picture theory relies on a specific notion of truth which is conceived of as the result of the application of a calculus of truth functions. By rejecting the picture theory, Wittgenstein also undermines the Tractarian notion of truth which is based on a comparison of language with reality.[36] Wittgenstein's later conception of meaning, however, leads to a different understanding of the notion. It is related to the meaning and understanding of phrases like 'is true' and its cognates in social contexts. For example, we have learnt in school that the calculation $12 \times 12 = 144$ is true and it is natural for us to call the calculation 'true' as *this* is how we have learnt to calculate.

Again, some forms of relativism might be worrisome in this regard as was pointed out earlier. According to Wittgenstein, different instructions might 'effect a quite different understanding'. Again, this does not entail, however, that the way we do mathematics is completely arbitrary. The way we have learnt to do mathematics does certain work for us and for the purposes that we want to achieve.[37] When we see inconsistencies in the framework of mathematical applications we either find new ways to do mathematics or we adapt our ends. As was argued, our language-involving activities have meaning under certain circumstances. We might imagine, for example, a world in which objects do not retain their degree of stability or their identity for a sufficient length of time. In such a world the rules that govern our use of simple mathematical propositions seem completely useless and would hardly find agreement amongst peers.

The idea that meaning is related to use and usefulness implies the possibility of different uses and applications that are considered as useful for particular purposes under certain circumstances. This is an important point for a proper understanding of what Wittgenstein calls a 'system' in his remarks in *On Certainty*. Notice that in these remarks Wittgenstein applies the concept of a system to quite different contexts. He speaks of 'the system of convictions', 'thinking within our system', 'our whole system of physics', doubts that 'form a system', 'our system of assumption', 'the system of our empirical propositions', 'the system of our empirical judgments', 'a system of what is believed', 'our whole system of evidence', 'our whole system of verification', 'our system of knowledge', 'the entire system of our language-games'.[38] Notice also that, in light of what was said earlier, for Wittgenstein the notion of system is not a theoretical concept that he uses for the purpose of explanation. He rather uses the notion to describe the circumstances under which we talk about different abilities of human beings: having convictions, having thoughts, doing physics, having doubts, forming assumptions, empirical judgements and beliefs, and so on. Arguably, it is not Wittgenstein's point to claim that those abilities and practices lead to different conceptual systems although he is also concerned with questions regarding the doxastic differences between human beings that are based on those abilities. As I understand Wittgenstein's remarks, the concept of a system rather emphasizes the integrative character that those abilities and practices have for our understanding, i.e. our understanding of the situation in which we find ourselves in reality.

I have already rejected the claim that our abilities and practices produce a fragmentation of different systems. The view that there might be incommensurable systems because there are a range of different uses to which language is put, misunderstands what the concept of a system does. For what it does not is to *explain* doxastic differences between human beings. This being the case, it is thereby overlooked that Wittgenstein's notion of a system implies an *ethical* solution to those differences. Independently of the question whether our system differs from the system that is formed by others, it is an ethical issue whether an agent is willing to adapt his or her system by being instructed into, and by learning new ways of using language.[39] This is one of the reasons why the 'duck-rabbit' example is particularly misleading in order to argue for the incommensurability of different systems. Perhaps an agent might be able to see only the duck, only the rabbit or neither of them. In any case the agent makes a choice to engage, or not to engage, with a certain training situation that might enable her or him by means of particular instructions to see what the agent was not able to see before. The claim that systems are incommensurable is a misinterpretation of the fact that agents might have not yet reached an agreement regarding the use of particular signs and expressions under certain circumstances. This is not a theoretical insight.

2.6 CSR and Wittgenstein's later philosophy

Understanding the later Wittgenstein as proposed, suggests there is a different interpretation one might bring to the basic assumptions of CSR, mentioned in the introduction. This does not mean that the aims of those assumptions must be ultimately rejected. It is without question that cognitive science, in its short history, has produced many interesting epistemic achievements. What I am here questioning, however, is the conceptual framework of the basic assumptions that underlie those epistemic achievements and of the methods that are applied to accomplish them.

It is clear by now that Wittgenstein's later philosophy does not provide cover for the idea of scientific explanations as social constructions in CSR. Scientists have to find agreement amongst their peers regarding the conceptual framework of the theories they propose. It is a misunderstanding, as I have argued, to assume that the possibility of such agreement is warranted by an independent reality that is related to the notion of objective truth. Rather, our understanding of what we call 'being true' is already informed by our practices and language-involving activities that are performed in reality. What enables scientists to generalize from one culture to another and from one time to another, to pick up on Trigg's train of thought, is that they come to an agreement about what they think should be considered as similar or different under the circumstances of the relevant issues. Based on such agreement they are able, as Trigg puts it, 'to predict from findings in Western society what could be the truth in China'.[40]

Proponents of CSR, however, must be careful about what they take themselves to be assuming with their basic assumptions. The assumption that empirical findings lead to objectivity and objective truth due to an independent reality is related to the kind of philosophical realism that Wittgenstein rejects. Yet the weaker version of the idea that empirical findings lead to objectivity and objective truth which is not necessarily committed to the assumption of an independent reality is related to forms of empiricism, idealism or solipsism. According to this version, the only way of achieving objective truth is by means of individual experience. Wittgenstein not only rejects this alternative but also prompts us to be cautious about falling into the trap of thinking that all sentences that appear to be empirical statements (i.e. they share a similar form with sentences that are used as empirical statements) are in fact used as such.[41]

Trigg rightly points out, that to opt for such a weaker version of objective truth paves the route to a reductionist conception of empirical science that is self-contradictory as it leaves us with no independent criteria to evaluate the objectivity of our experience. He holds instead that the philosophical assumption behind CSR is that 'how we think . . . is rather the result of the effect of the world and its problems on us, and a human understanding which has perhaps developed through evolution'.[42] He suggests that religious behaviour relies on common cognitive patterns and that empirical investigations of processes in the nervous system may result in an adequate understanding of these patterns. However, he remains rather silent on the philosophical question of how the results of those empirical investigations are related to our meaning and understanding. On the one hand, he accepts the insight that to identify cognitive patterns with physical processes is self-contradictory as the understanding of these processes relies, again, on certain cognitive patterns. On the other hand, if one accepts that the meaning of cognitive patterns is not determined by physical processes in the brain then no conclusions can be drawn from certain empirical results in cognitive science in terms of the meaning of the cognitive patterns in question.

It seems as if Trigg accepts this dilemma and responds by suggesting a carefully minimized role for CSR: 'It can enlighten us about human tendencies and dispositions and demonstrate the universality of impulses underlying religion'.[43] The interpretation of his proposal, however, still leaves us with the challenge of finding an alternative that avoids both philosophical realism and a form of empiricism. On his account scientists either use methods of empirical investigation to demonstrate the universality (threatened by a form of empiricism) or presuppose universality that underlies a certain understanding of the impulses that relates them to religion (under influence of some kind of philosophical realism).

Wittgenstein's later philosophy presents an answer to the challenge suggesting the rejection of empiricism in favour of a modified notion of realism.[44] His solution is based on the insight that the meaning and justification, for example, of the notion of reality and the notion of objective truth, do not depend on a philosophical theory. Much philosophical work was needed for Wittgenstein in the 1930s to arrive at this insight. During his work he develops an understanding and appreciation of reality that prevents a radical form of scepticism. Wittgenstein's later philosophy never comes at the cost of losing the notion of objective truth. It rather demands a change in the conceptual framework that underlies the use of this notion of truth: a change that does not entail idealism, constructivism, conceptual relativism or meaning-scepticism. Such changes in the conceptual framework that might modify the understanding of the central notions and basic assumptions of CSR are, however, yet to come.

Notes

1 See Gardner (1985), Bruner (1990).
2 See Xygalatas and McKay (2013).
3 See White (2021), especially Chapter 1.
4 Cf. PI, preface.
5 See, for example, Hintikka and Hintikka (1986), Kienzler (1997), Stern (2006), Engelmann (2013).
6 See *Tractatus* 2.1512 and 4.01.
7 See Wittgenstein 1979, 63. For an analysis of Wittgenstein's change in thought see Stern (1991).
8 See *Tractatus* 2.1513.
9 See Wittgenstein 1974, 213; cf. *PI* 141.
10 Cf. Wittgenstein 1975, 59. For deeper analyses of this account see Figueiredo 2019 and 2021.
11 Cf. *PI* 6. In this sense one might say that ostensive definitions establish connections *in language* (not between language and reality) by means of samples (e.g. actions, words and reactions). Samples are means of representation in language (e.g. to make colour statements; cf. *PI* 50), they do not establish a link between language and reality.
12 The training situation is a helpful example to demonstrate that reality is involved in the production of understanding by means of its incorporation in a particular kind of instruction. In this regard Wittgenstein points out that with different instruction 'the same ostensive teaching of [certain] words would have effected a quite different understanding' (cf. *PI* 6).
13 Cf. Wittgenstein 1969 (hereafter: *OC*), §559. See also *PI* §373: 'Grammar tells what kind of object anything is.' Both are also cited in Trigg's chapter.
14 Cf. Wittgenstein's quotation from Augustines *Confessions* I, 8 in PI 1.
15 Cf. PI 2.
16 Cf. PI 15.
17 Cf. *PI* 185.
18 Frege who saw the threat of psychologism in logic and mathematics put it somewhat stronger in his *Grundgesetze* by saying that 'here we have a hitherto unknown kind of insanity'. Cf. Frege (1893, XVI).
19 As Wittgenstein points out elsewhere, if no course of action can be determined by a rule because every course of action can be brought into accord with the rule (e.g. one might understand the teacher's order as order to add two up to 1,000, four up to 2,000, six up to 3,000 etc.), then it 'can also be brought into conflict with it. And so there would be neither accord nor conflict here'; cf. *PI* 201.
20 Cf. *OC* §§45, 47.

21 See *PI* 12. An earlier version of this remark can be found in an entry in Wittgenstein's manuscripts dated 12 January 1930 (Ms-107, 231–2); cf. Wittgenstein Archives at the University of Bergen (2016-).
22 One might also have reservations to call it a 'sign' as one could argue that the expression might have certain post-modern connotations in relation to semiotics.
23 Trigg's reference to the 'duck-rabbit' as an illustrating example seems an unfair choice in order to make the argument for incommensurability. Wittgenstein is arguing for the view that signs can be used for different purposes and nobody denies that, in general, certain uses exclude each other such that they cannot be applied at the same time for different purposes, for example, a screwdriver can be used to put a screw into the wall or off the wall; the order 'close the door!' can be given to shut the door from inside or outside the room; 'turn around!' can mean that the person should turn 180 degrees left or right, etc. The 'duck-rabbit' falls into this list as it can be used to demonstrate a duck or a rabbit but not both at the same time. This, however, does not allow for the claim that different uses *necessarily* lead to their incommensurability.
24 See PI 143–84. For discussion see Baker and Hacker 2005b, 313–56.
25 Wittgenstein points out that 'it is in their *language* that human beings agree. This is agreement not in opinions, but rather in form of life'; cf. *PI* 241 (also cited by Trigg).
26 *PI* 242.
27 In fact, Wittgenstein compares the situation in which the pupil suggests his way to use '+2' with the situation in which 'it comes naturally to a person to react to the gesture of pointing with the hand by looking in the direction from fingertip to wrist, rather than from wrist to fingertip'; cf. *PI* 185. As I argue, this should not be read as remark about correct and incorrected applications but a remark about what might motivate our choices (e.g. a natural reaction) to suggest certain applications to interacting peers.
28 See for example Boghossian 2006, Glock 2008.
29 Trigg, (Chapter 1, p. 16). I refer to Wittgenstein's notion of a system in the next section. It is perhaps also noteworthy that the version of philosophical realism in the *Tractatus* is rather different to what philosophical realists before Wittgenstein adhered to.
30 Cf. *PI* 242; also cited earlier.
31 Cf. *OC* §§608–12 and Figueiredo 2015 for discussion. Again, the 'duck-rabbit' example can be in certain ways misleading here. Of course, we might imagine a disagreement between agents who only see a duck in the picture and those who only see a rabbit. This does not exclude the possibility, however, that someone who only sees the duck might be instructed to also see the rabbit. Attempts might be not successful but chances of success certainly exist and thus success is not impossible.
32 Cf. *PI* 415. Trigg draws heavily on this point suggesting that what Wittgenstein is referring to might be the scope of empirical studies in CSR. I will come back to this point later.
33 Crary 2005 points out that it is often suggested that Wittgenstein repudiates a classical theory of meaning in favour of a 'use-theory of meaning' which has relativistic implications.
34 This view seems in agreement with Baker and Hacker's view: '[G]rammar can be said to be "responsive to reality" in the following sense: that were the world different in specifiable ways, certain features of our form of representation would no longer be *useful*. So too, if human nature were different, parts of our grammar might no longer be *usable*. And if we had a different grammar we would say and do quite different things'; Baker and Hacker 2009, 339.

35 This in itself refers to the ability to understand things in a different way, to change one's view of certain concepts (e.g. reality) and to use them differently which means to give them a different meaning. It will provide great difficulties for someone who is convinced that there is only one meaning for a concept to accept a different view. As Wittgenstein suggested already in January 1930: 'Strangely enough, the problem of *understanding* language is connected with the problem of the Will'; cf. Wittgenstein 1975, 58 (see also Ms-107, 231).

36 Cf. *Philosophical Investigations*, §§134–42. It should be pointed out that the *Tractatus* distinguishes between the truth of logical and empirical statements. The truth of logical statements is based only on their logical structure and on the meaning of the terms. Logical statements are true under all conceivable circumstances; thus their truth is independent of the contingent facts of the world (cf. Carnap (1963); he describes this as Wittgenstein's 'most important insight'). The truth of empirical statements is determined by the application of a calculus of truth functions under certain truth conditions (cf. *Tractatus* 4.063).

37 Cf. Wittgenstein 1978, 37–8.

38 Cf. *OC* §§102, 104, 108, 126, 134, 136, 137, 141, 142, 144, 185, 279, 285, 327, 410, 411.

39 Alice Crary's interpretation seems convincing that reads Wittgenstein in a sense in which he suggests 'a view of ethics as concerned with a dimension of all discourse [i.e.] the idea that all thought has an ethical dimension'; cf. Crary 2005, 294–5. Compare this also to Wittgenstein's remark that 'work on philosophy is actually closer to working on oneself. On one's own understanding. On the way one sees things. (And on what one demands of them)'; cf. Wittgenstein 2005, 300.

40 Cf. Trigg, 20.

41 See, for example, Wittgenstein's remarks about what might be the meaning of 'believing in the Last Judgment'; cf. Wittgenstein 1967.

42 Trigg, 20–21.

43 Trigg, 23.

44 Cf. Wittgenstein's famous remark: 'Not empiricism and yet realism in philosophy, that is the hardest thing'; cf. Wittgenstein 1978, 325. It is noteworthy that Wittgenstein relates this remark to positions in philosophy whereas I have related it to scientific problems.

3

Pascal Boyer's miscellany of homunculi

A Wittgensteinian critique of *Religion Explained*

Robert Vinten[1]

3.1 Boyer's understanding of religion

In *Religion Explained* Pascal Boyer uses a conceptual framework that centres around what he calls 'inference systems'. An inference system, according to Boyer, is a system in the brain that produces inferences when it is activated by things we perceive in our environment. So, for example, human beings have what he calls an 'intuitive physics' system. Things we perceive, such as a billiard ball hitting another billiard ball, are 'input' for the system. Given this input, the system infers that something invisible is present, that is, a cause.[2] Other inference systems include the 'person file system' (a system in our brains that keeps track of who people are), the intuitive psychology system (which makes inferences about what other people are feeling and thinking) and the contagion system (which makes inferences about how you should respond to an object deemed to be potentially poisonous or infectious).[3] These 'inference systems' are made to do a lot of work in explaining religion in his book. Boyer claims that 'the activation of a panoply of systems in the mind explains the very existence of religious concepts *and* their cultural success *and* the fact that people find them plausible *and* the fact that not everyone finds them so *and* the way religion appeared in human history *and* its persistence in the context of modern science'[4]. He claims that these systems make inferences without *us* being aware of it and that they also exchange information with each other 'behind the scenes'.

Boyer sees what he is doing in *Religion Explained* as being a scientific endeavour which unites insights from evolutionary theory, neurophysiology, psychology and anthropology. The cognitive science of religion, of which Boyer is one of the principal proponents, is still a relatively young discipline; approximately thirty years old,[5] and it relies upon evidence from psychologists and cognitive scientists that have been produced in recent decades. This experimental data, Boyer thinks, is overturning older explanations of religion which explain its appeal in terms of the comfort it provides, the explanatory power it apparently affords, or its role in propping up a particular social order.[6] All of these earlier explanations contain a germ of truth according to

Boyer but all are also insufficient and misunderstand their object. On Boyer's account, our understanding of religion becomes sharper when we look carefully at how our minds process sensory input.

His book is subtitled *The Evolutionary Origins of Religious Thought* but it is not principally concerned with working out when it was that religious notions appeared in the course of evolution (although he does have a few things to say about that). What Boyer wants to work out is why it is that religious notions appeal to human beings and how it is that they are passed on successfully between human beings. Boyer thinks that looking to evolutionary theory helps to explain the appeal of religious notions and helps to explain why certain kinds of notions are passed on more successfully than others. When discussing religious notions or concepts, Boyer sometimes uses the term 'memes', which he describes as '"copy-me" programs, like genes'.[7] These memes, like genes, are replicated, according to Boyer, and different people end up with slightly different versions of these memes as the replication process goes on. Memes are passed between individuals and communities according to their fitness for the human mind and for human life.[8]

The 'meme' and its transmission via a process modelled on human genetics is one biological metaphor, but Boyer also invokes another: the evolutionary explanation for people becoming religious is somewhat like the evolutionary explanation for people catching viruses. Evolving into a human does not mean that you will necessarily catch a cold but humans have evolved lungs such that they are susceptible to catching colds. Similarly, human beings have evolved to have brains such that they are susceptible to "catching" religion.[9] There is no simple, single explanation for this susceptibility, on Boyer's account. People 'catch' religion for various reasons, because they have evolved to have a variety of different inference systems that make them susceptible.

One way in which people are susceptible to 'catching' religion is that they have evolved in such a way that they are very sensitive to signs of agency in their surroundings. Most of human evolution has taken place with human beings living in circumstances such that they were prey to some predators and so had to avoid being predated and in circumstances in which humans were also predators themselves and needed to hunt prey. So, it was hugely important to human beings, throughout most of their evolutionary history, that they should be able to detect animate creatures in their environment – either to catch them or to avoid them. A human that was inclined to overdetect agency, to think that there was a creature around when there was not in fact a creature around, would be more successful than one that underdetected agency. Failing to detect a predator as it approaches you could easily be deadly but overdetecting agency does not have similarly high costs. In *Religion Explained* Boyer puts it like this:

> Our agency detection systems are biased towards overdetection . . . it is far more advantageous to overdetect agency than to underdetect it. The expense of false positives (seeing agents where there are none) is minimal, if we can abandon these misguided intuitions quickly. In contrast, the cost of not detecting agents when they are actually around (either predator or prey) could be very high.[10]

As a result of us being geared towards perceiving things as agents and seeing things in terms of volitional behaviour we concoct agents that do not really exist (gods, ghosts, spirits, witches etc.). It is worth noting here that Boyer's explanation is not put in terms of human beings being inclined to think they have perceived agents where there are not in fact any agents. Boyer's explanation is in terms of a system in the brain, the 'agency detection system' (one of our inference systems) which has evolved to detect living things in the environment.[11] That is, due to the action of these unconscious systems, certain ideas are easily manufactured or taken up by the human mind.

The inferences made by our inference systems are made without the person in question being aware of them, according to Boyer. Inference systems 'run so smoothly in the mind that we are not aware of their operation . . . our minds run . . . chains of inferences automatically, and only their results are spelled out for conscious inspection'.[12] When discussing inference systems Boyer sometimes describes them as being in the *mind* (as in the quote from him in the previous sentence) and sometimes describes them as being in the *brain.*[13] For example, he says that 'the brain's inference systems produce explanations without our being aware of it'.[14] According to Boyer these systems in the mind/brain do all kinds of things without us being aware of it: 'whatever you see is perceived, interpreted, explained, recorded by the various inference systems'.[15] Our inference systems also *understand* various things, such as the physics of solid objects and physical causation,[16] try to figure out how to handle artefacts,[17] produce descriptions of states in other people's minds[18] and produce representations of what is going on inside people's heads.[19] So, inference systems can produce explanations, perceive things, interpret what you see, record what you see, understand physics, figure things out and produce descriptions and representations of other people's mental states. The list of things that inference systems can do is longer than this. Inference systems are also said to imagine making gestures,[20] make assumptions,[21] keep track of the number of objects behind a screen,[22] compute whether other people are trustworthy,[23] pay attention to eye contact,[24] treat pets as if they are people,[25] plan courses of action and assign value to those courses of action,[26] perform precise calculations,[27] take information from the optic nerve[28] and various other things.[29]

The explanation of religion in terms of the agency detection system is not the only explanation of religious belief provided by Boyer. One of the ways in which he explains religion is in terms of particular religious concepts being memorable. Good religious concepts: ones that will be passed on and remembered are ones that are made up of an ontological label and a tag that is attached to the label that contradicts an element of the label. An ontological label is a high-level category (a category that incorporates a lot of members and so supplies us with a lot of inferences to make). For example, at a low level we might consider a particular thing, my cat Bósforo. Bósforo is a particular example of something belonging to the category 'cat' and if I am in possession of the category 'cat' then I can infer all kinds of things about Bósforo when I am aware that he belongs to that category (e.g. that Bósforo will likely enjoy eating meat and fish but not vegetables, that Bósforo will breed with other cats but not with animals from other species – because these are typical of cats). At an even higher level than the category 'cat' is the category ANIMAL and it is at this level that Boyer uses the expression 'ontological label'. Examples of ontological labels include PERSON, TOOL, NATURAL OBJECT,

ANIMAL and PLANT.[30] We have just seen that someone's possession of the category or concept 'cat', on Boyer's picture, is associated with various pieces of information, such as that cats breed with other cats and not with other animals. Having this information tied to our concepts means that we are able to infer all kinds of things about the things we perceive. When we get to a higher level, to the level of the ontological labels, we have even more information associated with those labels. Possession of a 'label' like ANIMAL enables us (or systems in our brains) to make a lot of inferences. What is interesting about religious concepts is that they commonly involve one or another of the five ontological labels listed earlier but they also contain 'information' tagged to them which contradicts something that we would normally infer from the ontological label. So, for example, upon seeing a person our inference system would typically infer (or assume) that the person is in control of their own actions. One of the bits of information we have tagged to the ontological label PERSON is that a person is something that is aware of its surroundings and can move itself around in them at will. But when it comes to the religious concept ZOMBIE what we have is a concept that is made up of the ontological label PERSON (zombies are a kind of person) but tagged to that label is the information that a zombie does *not* have control of its own actions and so 'religious' concepts like ZOMBIE are 'counterintuitive'. Boyer lists various other central religious concepts and attempts to show that they fit this model of involving an ontological label and a special 'tag'. So the concept of 'an omniscient God', according to Boyer, is made up of the label PERSON and the tag 'special cognitive powers'; the concept of 'visiting ghosts' is made up of the label PERSON and the tag 'no material body' (contradicting our expectation that a person should have a body); and the concept of 'reincarnation' is made up of the label PERSON and the tags 'no death' and 'extra body available'. Because each of the labels is associated with a lot of information Boyer describes them as 'minitheories'. According to Boyer, research has shown that counterintuitive concepts like those found in religion are recalled well[31] and especially if they include just one 'violation' (i.e. just one tag that contains information conflicting with our expectations, given the ontological label).[32] It is important for something to work well as a religious concept that although it contains a counterintuitive element in other respects the concept functions in much the same way as the label. So, for example, although ghosts are counterintuitive in not having a body, in other respects they are very much like people. They can communicate, they can walk, they can perceive things in their environment and so on.[33]

Boyer's explanation of religion involves more than just the overdetection of agency by our agency detection systems and the appeal of counterintuitive concepts to systems in our brains. For example, Boyer explains the close association between religion and death in terms of the fact that death activates the contagion system (a system in the brain that perceives something to be contaminated and makes appropriate inferences) but although the contagion system produces 'precise intuitions' about contamination, the notions about death that are available to us at the conscious level (remember that the work of inference systems goes on without us being aware of it) are less precise and so more apt to be taken into a religious system of concepts (e.g. notions of 'bad air' or 'impurity')[34]. The predation system in our brain that has evolved during a time in which human beings had to be sharply aware of both predators and prey is also activated by death because death is the upshot of successful predation. The animacy system represents a dead body to us as something inanimate but our person-file system still recognizes a dead body as a person and so the mixture of these various systems gives death a sense of strangeness and produces feelings of guilt. At a funeral people feel guilty about burying a relative, according to Boyer, because '[d]isposing of the corpse is mandated by some mental systems for which this makes sense because the body is represented as an inanimate object (animacy system) and as a signal of danger (contagion system, predation system). But disposing of the corpse also involves handling a person who is *not yet* absent, as far as our person-file system is concerned'.[35]

At the start of *Religion Explained* Boyer said that in order to understand religion what we need is a 'precise description of how brains receive and organise information'.[36] As he goes on, we have seen that he develops what he takes to be a more precise description of how brains receive and organize information in terms of a variety of inference systems. Each of the various explanations that he gives – of how religious notions are successfully passed on, of why we feel guilty at funerals, of why people organize purification rituals around a death and so on – are given in terms of how particular inference systems are activated by things in the environment (agents, dead bodies, etc.) to produce expectations and inferences in those systems.

3.2 – A Wittgensteinian critique of Boyer's conceptual framework

As we have seen, Boyer's investigation is an attempt to give 'a precise description of how brains receive and organise information'[37] and Boyer spells this out in terms of the activities of various inferences systems in the brain which, according to Boyer, produce explanations, make assumptions, perceive things, attempt to figure out how to handle artefacts, describe other people's mental states, compute whether other people are trustworthy, perform calculations, discriminate between the dead and the living and even treat pets as if they are people; all without us being aware of it. The fact that we are unaware of the activities of these inference systems, in Boyer's view, shows the need for the work of cognitive scientists to uncover what is going on in people's minds or brains (Boyer uses 'mind' and 'brain' interchangeably). Boyer tells us that it is not just speculation that there are these inference systems. That there are inference systems in the brain 'is made manifest by both neuro-imaging and pathology'.[38]

However, before we take Boyer at his word, we should first be careful to ensure that we know what taking him at his word involves. Can we make sense of the claim that neuroimaging reveals systems in the brain that make inferences, form assumptions and present us with explanations? A remark from the philosopher Ludwig Wittgenstein can help us to get clearer about this. In his *Philosophical Investigations* Wittgenstein observed that 'only of a living human being and what resembles (behaves like) a living human being can one say: it has sensations; it sees; is blind; hears; is deaf; is conscious or unconscious'.[39] Wittgenstein's observation is an observation about what it makes sense to say or about the grammar of psychological expressions. So, for example, it makes sense for me to say, 'I [a human being] am a bit deaf in one ear' but it does not make sense to say, 'My brain is a bit deaf in one ear'. After all, one hears with one's ears and brains do not have ears. Even if we say (correctly) that somebody's deafness is due to brain damage we nonetheless say that it is the man or woman that is deaf, not their brain. In order to hear something I might bring my ear closer to the sound or turn towards it. I might cup my hand behind my ear to hear more clearly. But there is no equivalent in the case of brains. Watching a boxing match between the heavyweights Deontay Wilder and Tyson Fury I might see Wilder hit Fury hard, Fury fall to the floor and be amazed that Wilder has knocked Fury out; that Fury is *unconscious* as a result of being punched by Wilder. I can see that Fury (a human being) is unconscious because he is lying unmoving and is completely unresponsive to stimuli in his environment. The referee waves his hands in front of Fury's eyes and there is no movement from Fury. It makes sense to say that 'Tyson Fury is unconscious' but it does not make sense to say that Tyson Fury's brain is unconscious. After all it was not Tyson Fury's brain that was punched and that fell to the floor and was unresponsive to the referee's waving, it was Fury. If I walk into the kitchen and open the cupboard containing my cat's food and my cat follows me in and rubs himself up against my leg I might say that 'Bósforo saw me going towards his food and got excited' but not that 'Bósforo's brain saw me going towards its food and got excited.' His brain did not rub against my leg, cannot see and does not eat food and we cannot make any sense of talking about a brain being excited at the prospect of eating food.[40]

Boyer himself seems to acknowledge this point when he is discussing the way that psychological terms are attributed to groups: 'Villages do not "perceive" situations, committees cannot "remember" what happened, companies have no particular "desires", simply because all these groups are not persons.'[41] So, it seems as though Boyer thinks that *persons* are appropriate subjects to apply psychological expressions to.[42] However, Boyer also clearly thinks that psychological expressions can legitimately be applied to brains, minds and parts of brains, although we cannot apply them to groups. Yet this gets things the wrong way around. It is perfectly possible to apply psychological terms to groups. It makes sense to say that 'the leadership of the Conservative Party *believe* that Britain is better off outside of the EU', that 'M. Klein & Company *intend* to acquire an additional $50 million worth of Churchill III's shares', that 'we, the undersigned, are *shocked* and *disgusted* to read the evidence in the report', or that 'the schoolchildren *saw* the dolphin leap out of the water'. It is possible to apply psychological expressions to groups because groups can in some ways exhibit the kind of behaviour that living human beings do. They can issue statements, they can write letters and they can act in pursuit of goals. The Wittgensteinian philosopher Peter Hacker uses the term 'artificial persons' and gives 'a company or institution, a government or a state' as examples in his recent book *The Passions*.[43] However, brains cannot behave like persons at all. There is no such thing as a brain issuing a statement, writing a letter or acting in pursuit of goals. Brains cannot engage in verbal or non-verbal behaviour in the way that human beings do. They do not have mouths with which to speak, faces on which you might read expressions or bodies which they might move around in response to things in their environment. A brain's environment is the inside of a person's head. We do not know what would be involved in asking a brain a question or how a brain might answer one but we do know what it is like for human beings and artificial persons, such as companies, to respond to queries. We know what it means to say 'John saw the dolphin' and we know what it means to say 'the class saw the dolphin' but Boyer's expressions are grammatically abnormal and cause confusion. We do not know what it means to say 'the brain saw . . . ' or 'the brain infers . . . '.[44]

So, why does Boyer speak in this way about inference systems, brains or minds believing, perceiving, assuming and planning? One possibility is that he is using these familiar expressions in an unfamiliar way, as some kind of technical terms from cognitive science. If that is so then it is strange that Boyer does not explain the meanings of the expressions as they are applied to brains, parts of brains or minds. Another possibility is that Boyer is talking about brains as a shorthand way of talking about what is going on in brains when *the person* (or the human being) in question is perceiving something, assuming something or planning something. But this cannot be right because it is clear that Boyer does not want to ascribe the relevant terms to the person when he is talking about the activities of the inference systems. As noted earlier, the inference systems do all of their perceiving, assuming, inferring and planning without *us* being aware of it. This is essential to Boyer's 'theory': that the religious person makes bad inferences and forms false beliefs without being aware of them. It is because the processes of 'inference systems' are hidden that 'scientific analysis' is needed to uncover what is going on. That an inference system is inferring something, on Boyer's account, clearly does not mean that *the person* in question is inferring something. I suspect that the reason why Boyer speaks in the way that he does, in attributing psychological predicates to inference systems, lies in the fact that Boyer has adopted a Cartesian framework more generally in his understanding of the mental lives of human beings. His model of perception, his understanding of the application of psychological terms to other people (involving a 'theory of mind'), his account of concepts as 'minitheories' and his understanding of what he calls 'intuitive physics' all show that he adheres to a model of our mental lives where inferences or computations are seen to be necessary where in fact they are not necessary.

To give one example of where Boyer takes a wrong turn; in discussing how it is that human beings understand the actions of other people he explains that the brain/inference system has to infer that other people are thinking certain things because 'thoughts are invisible. You cannot observe them directly, you have to infer them'.[45] There are echoes here of Descartes's division of the human being into an immaterial substance and a material body that are causally connected via the pineal gland. Although Boyer does not subscribe to this model and does not talk in terms of an immaterial substance, he does retain the idea of something invisible hidden behind the bodies that we (or our brains) perceive. The model is one on which in coming to understand what is going on with other people all we have to go on is what we can perceive, but mental states and processes cannot be perceived and so it must be that we infer them from what we can perceive. Like Descartes, Boyer ascribes mental states, processes, and acts to the mind (which, in Boyer's case, is used interchangeably with 'brain'). Boyer's adoption of this model also goes some way to explaining why he thinks of what he is doing as theoretical. The things we are interested in when we want to investigate religion: beliefs, thoughts and so on, are taken to be invisible things, like atoms. Work has to be done in uncovering what is going on in people's mental lives because we have to observe the 'surface' (movements of bodies) to uncover what is underneath. On Boyer's picture there is a strict division between 'inner' and 'outer'. In fact, there is a division in at least two stages because even in perceiving something 'outer'; the behaviour of a body, the brain must make inferences about what is 'out there' based on the information that it receives. So, there are inferences involved in perception, the upshot of which provides the material for inferring what is going on in the minds of the people around us. Inferences are ubiquitous in the world as Boyer understands it. Whenever we perceive things and particularly whenever we perceive a living creature, there have to be inferences going on all of the time. But given that we are not aware of continual inferences being made it must be that the inferences are being made by something other than the human beings walking around looking at things and talking to people.

But Boyer's account of the relationship between the 'inner' (mental/invisible) world and the 'outer' (perceptible) world of material things is awry. As Wittgenstein said in his 'Lectures on Private Experience and Sense Data',[46] 'We must get clearer about how the metaphor of revealing (outside and inside) is actually applied by us; otherwise we shall be tempted to look for an inside behind that which in the metaphor is the inside.'[47] So, let's look at a few cases in which the grammar of 'inner' and 'outer' is clear and relatively unproblematic. (i) Imagine that you are at a party with a friend and you spot a mutual acquaintance. Noticing that your friend has not seen your mutual acquaintance you decide not to reveal the fact that the mutual acquaintance is there because you would rather not go over and talk to them at the time. It is possible that a bit later in the party you might tell your friend that your mutual acquaintance is there and that you might then go over and talk to the person in question. In that case you would have revealed that you have seen the mutual acquaintance and now you and your friend would both be aware that the mutual acquaintance is there. Your friend would be able to see exactly what you have seen. What was concealed has now been revealed. (ii) Imagine that a military leader is plotting to overthrow the government and they conceal their plans for the overthrow of the government using a code. The military leader shares their thoughts with a select group by writing them down in the code and the group knows how to use the code but other people are completely unaware of the plot. People outside of the select group do not have access to the thoughts of the military leader or to the military leader's plans to overthrow the government. But then imagine that somebody from outside of the group obtains the documents and is able to crack the code. They might then announce to the government what they have discovered and the plan might become public knowledge and the military leader could well be condemned as a result of the discovery. What was *hidden* has been *revealed* and the *revelation* forms the basis for action against the military leader. (iii) it is possible for somebody to be in pain and not reveal it. This is common enough. They might not give any hint of being in pain in their expressions, they might not say anything about the fact that they are in pain, and they might take pain medication when other people are not around. But it is also possible that somebody might stub their toe against a door and yell 'ow f%&k! My toe!' and when somebody asks what is going on they can say, while wincing, that they have stubbed their toe and that they are in terrible pain. In that case, the person has revealed that they are in pain. It is manifest to onlookers.[48]

In cases like these it is not available to us to witness one of these moments of revelation (that the mutual acquaintance is at the party, that the military leader is plotting to overthrow the government, that somebody has stubbed their toe) and to then say, 'Well, that is only behaviour, the mental state in question is still hidden from us'. In that case we are not 'tempted to look for an inside behind that which in the metaphor is the inside' as Wittgenstein put it. Another thing to notice about these cases is that we do not think about 'concealment' in terms of 'keeping in the realm of the inner'. A coded diary in which thoughts are concealed is a publicly available physical object that anyone with functioning eyes in good light can see, although they might not be able to gain access to the diary or to crack the code in which it is written.

These cases suggest that Boyer's way of thinking about our mental life, with a clear distinction between an invisible 'inner' realm of thought and an outer realm that is perceptible, is confused. According to Boyer's model it would make sense to say that we have to infer something hidden/invisible, such as the military leader's thoughts, even after the moment of revelation upon cracking the code and reading the military leader's documents. There is something hidden away beneath the surface even as the documents spell out very clearly a plan for the overthrow of the government. On Boyer's view, of course, it is not only the case that I have trouble inferring what others believe but also that I often do not know what I, myself, believe, because my beliefs are formed by brain mechanisms without me being aware of it.

But these examples might not be enough to sway somebody like Boyer. There is still the temptation to say that even though we ordinarily speak in these ways that there are nonetheless inferences involved to get at something inner when we look at things in a more strictly scientific manner. We should perhaps acknowledge that our ordinary way of speaking does not get at how things are. We need to be careful in thinking about the cases to think about all of the possibilities. For one thing, if somebody conceals their thoughts in a diary it remains possible that the thoughts in the diary are not the thoughts in the person's head. It is possible that somebody might place a diary somewhere that they know it will be found in order to *deceive* someone. It is possible in cases where somebody thinks they have seen someone (e.g. a mutual acquaintance) that they might have misperceived and have been mistaken about whether the person in question was an acquaintance of theirs. It is possible for someone to act as though they are in pain when they in fact are not, in order to garner sympathy. These possibilities of deceit, misperception and of acting lead us to think that there is a 'gap' that has to be bridged and inferences provide that bridge.

However, these temptations should not be succumbed to. Although it is true that people sometimes deceive, that they sometimes mistakenly think they have seen something when they haven't and that they sometimes pretend to be in pain, these considerations do not imply that another step must be involved. To think so is to misconstrue the logic of our psychological concepts. The fact that it is possible to be mistaken about what we have perceived certainly does not imply that what we perceive is something mental and that our brain needs to infer from 'mental representations', 'ideas' or 'sense data' what is really going on 'out there'. The fact that deceit is possible does not imply that we must always infer someone's motives or thoughts, and the fact that people sometimes pretend to be in pain does not imply that we must always infer whether people are in pain or not from their behaviour.[49]

There is a temptation to think that because misperception, deceit and acting are possible that pieces of behaviour do not imply that someone is in a given mental state or undergoing a certain mental process and so there is no *logical* relationship between somebody's behaviour and their mental life. But it is possible to acknowledge that people can be deceitful and that they can sometimes act in a way that disguises their thoughts or their pains and retain the claim that there are logical relationships between behaviour and mental goings-on. It could be that a certain behaviour is a (defeasible) criterion for somebody being in a certain mental state. This is not to say that, *contra* Boyer, it is straightforward to understand the emotions and thoughts of others or our own emotions and thoughts. People do sometimes disguise their emotions and people can be difficult to 'read' even if there is no disguising, deception or acting going on. The behavioural manifestations of pain, grief and joy and the associated facial expressions differ from person to person. The grammar of psychological expressions is complex and it is this complexity is one of the things that has led philosophers, psychologists and cognitive scientists astray in trying to understand human psychology.[50]

What does it mean to say that there are logical connections between behaviour and our mental life? On Boyer's account, where there is no logical/conceptual connection between pain and pain behaviour, it remains a possibility that all the people who smile and behave in a happy, easy-going way, are in dire pain, while all the people who say they are in pain, who cry 'ow!' loudly, and who wince after injuring themselves are not in pain. But, as the Wittgensteinian philosopher Peter Hacker points out 'it is part of what we mean by "toothache" that it is exhibited in these forms of pain behaviour. We learn to say "he is in pain" when he behaves thus and his utterance "It hurts" is no less a criterion of pain than the groan'.[51] We can learn to pretend that we are in pain but it is worth remembering that it is *learned* behaviour. A baby cannot pretend to be in pain.[52] Saying that certain forms of behaviour are criteria for being in pain does not mean that they entail that someone is in pain – but entailment is not the only form of logical connection.[53] Those pieces of behaviour are defeasible criteria (they might be 'defeated' by some other piece of evidence – somebody who has complained of a terrible pain in their leg who is then seen to walk off with no indication of being in pain gives us reason to doubt that they are). If the criterion is not defeated then we can be certain that the person is in pain[54]. As Wittgenstein says 'We must not look for "toothache" as something independent of behaviour. We cannot say: "Here is toothache and here is behaviour – and we can put them together in any way we please."'[55]

Notes

1 I'd like to thank participants in the ERB (Epistemology of Religious Belief) workshops at Universidade Nova in Lisbon for their comments on an earlier draft of this chapter. Nuno Venturinha, Sofia Miguens, Alexandra Dias-Fortes, Modesto Gómez-Alonso, Thomas Carroll, Florian Franken Figueiredo, Guy Axtell, Hans Van Eyghen, E.

Thomas Lawson, and Gorazd Andrejč have all given me useful feedback. I would also like to thank Christopher Hoyt who has read more than one draft of this chapter and given me comments and advice. Of course, responsibility for any errors lies entirely with me. I would also like to thank the FCT (Fundação para a Ciência e a Tecnologia) for their financial support while I was working as a postdoctoral research fellow within the project 'Epistemology of Religious Belief: Wittgenstein, Grammar and the Contemporary World' (Ref: PTDC/FER-FIL/32203/2017).

2 Boyer 2001, 98–9.
3 Boyer 2001, 218–19.
4 Boyer 2001, 298.
5 E. Thomas Lawson and Robert N. McCauley are considered the founders of the discipline, with their book *Rethinking Religion*, 1990.
6 Boyer 2001, 5–31.
7 Boyer 2001, 35.
8 For an excellent recent critique of this conception of memes see pp. 488–91 of Bennett and Hacker (2022).
9 Boyer 2001, 4.
10 Boyer 2001, 145.
11 Boyer is not alone in trying to explain the spread of religion in terms of a part of the mind or brain that detects agency. He credits the idea to Justin Barrett and Barrett presents much the same kind of account of the spread of religion as Boyer does. Instead of using the expression 'inference system' Barrett speaks about 'mental tools', and these supposed 'mental tools' play much the same role in his account of religion as 'inference systems' do in Boyer's account. Our mental tools include an object detection device, an animal identifier, an object describer, a living thing describer, a social exchange regulator, a social status monitor and an agency detection device (Barrett 2004, 3–6). The Agency Detection Device he describes as 'A mental tool that determines whether an object is an agent or whether some event or trace is the result of agency. In many contexts, ADD may detect agency given only scant information; hence, it may also be termed the hypersensitive agency detection device' (2004, 125. See also Barrett 2000).
12 Boyer, 2001, 17.
13 The same sort of ambiguity is present in Justin Barrett's work. When he is introducing the notion of a 'mental tool' he asks us to think of 'the human mind as a workshop filled with racks of tools' but soon afterwards speaks of the brain as having 'specialized tools' (2004, 3).
14 Boyer 2001, 18.
15 Boyer 2001, 30.
16 Boyer 2001, 96–7.
17 Boyer 2001, 102.
18 Boyer 2001, 102.
19 Boyer 2001, 103.
20 Boyer 2001, 104.
21 Boyer 2001, 110.
22 Boyer 2001, 111.
23 Boyer 2001, 125.
24 Boyer 2001, 151.
25 Boyer 2001, 221.
26 Boyer 2001, 239.

27 Boyer 2001, 250.
28 Boyer 2001, 303.
29 There is a similarly long list of the things that 'mental tools' in Justin Barrett's work, including 'becoming confident in their nonreflective beliefs in God' (2004, 70).
30 Boyer says that there are very few of these high-level 'labels'. In fact, he thinks that 'ANIMAL, PERSON, TOOL . . . NATURAL OBJECT . . . and PLANT more or less exhaust the list.' (2001, 78).
31 According to Boyer, research has found that these concepts are both memorable and also passed on accurately (see Chapter 2 of *Religion Explained* 51–91).
32 Boyer 2001, 85.
33 Again, Justin Barrett presents a very similar account of how religious concepts spread to the one we find in Boyer's work. In his book *Why Would Anyone Believe in God?* Barrett agrees with Boyer that religious concepts spread because they are 'minimally counterintuitive'. Minimally counterintuitive concepts, 'concepts that . . . violate a small number of assumptions generated by the mental tools called categorizers and describers', are, according to Barrett 'quite memorable and easy to transmit to others and may also be believable' (30). Notice that there is an oddity here in describing concepts as the kind of thing that might be believed, supported (39) or affirmed (61). We might believe that something or other is the case or affirm that it is but we cannot believe a concept (we might acquire them, use them or understand them but we cannot *affirm* them. Concepts are not the kind of thing that can be supported, that can be true or false. As Wittgenstein notes, our concepts are not justified by reality, although *claims* about how things are obviously are justifiable by reference to how things are ('The connection between "language and reality" is made by definitions of words, and these belong to grammar, so that language remains self-contained and autonomous' (*Philosophical Grammar*, 97) 'Why don't I call cookery rules arbitrary, and why am I tempted to call the rules of grammar arbitrary? Because "cooking" is defined by its end, whereas "speaking" is not. That is why the use of language is in a certain sense autonomous, as cooking and washing are not. You cook badly if you are guided in your cooking by rules other than the right ones; but if you follow rules other than those of chess you are *playing another game*; and if you follow grammatical rules other than such-and-such ones, that does not mean you say something wrong, no, you are speaking of something else.' (*Zettel*, §320).
34 Boyer 2001, 215.
35 Boyer 2001, 219. See Christopher Hoyt's chapter in this volume for a detailed examination of Boyer's claims about our apprehension of death and perception of bodies.
36 Boyer 2001, 3.
37 Boyer 2001, 3.
38 Boyer 2001, 102.
39 Wittgenstein 2009, §281. Long before we were presented with this remark from Wittgenstein, Aristotle had said something similar: ' . . . to say that it is the soul [*psuchē*] which is angry is as if we were to say that it is the soul [*psuchē*] that weaves or builds houses. It is doubtless better to avoid saying that the soul [*psuchē*] pities or learns or thinks, and rather to say that it is the man who does this with his soul [*psuchē*]' (Aristotle, 1956, 408b12-14) and in *The Physical Basis of Mind* George Henry Lewes gave us a modern version of Aristotle's conception, saying that: 'It is the man and not the brain, that thinks; it is the organism as a whole and not one organ that feels and acts' (Lewes 1891, 498). Wittgenstein's remark has since inspired various

reformulations, including Anthony Kenny's 'homunculus fallacy' which says that it is a fallacy (or at least leads to fallacies) to ascribe psychological attributes to the brain. Given that we ascribe psychological attributes on the basis of behaviour (human behaviour or behaviour resembling it), it must be that if we ascribe psychological attributes to parts of the brain then the brain is made up of many homunculi – little things that resemble human beings in their behaviour. But then the question arises of how we are to explain the behaviour of the homunculi. Do they also have homunculi in them that explain their behaviour? (see Kenny 1971). Maxwell Bennett and P. M. S. Hacker prefer to speak of a 'mereological fallacy', which they say is the '. . . mistake of ascribing to the constituent *parts* of an animal attributes that logically apply only to the animal *as a whole*' (Bennett and Hacker 2022, 84).

40 Amongst those who agree that there is a problem here in attributing psychological attributes to brains that only properly apply to whole creatures (human beings, persons, and creatures that resemble human beings in their behaviour) there is disagreement about how best to characterize the problem. We have already seen in the earlier footnote that Anthony Kenny describes the problem in terms of a 'homunculus fallacy'. More recently, Maxwell Bennett and Peter Hacker have described the problem as being a 'mereological fallacy' (i.e. a fallacy, or a mistake that leads to fallacies, of ascribing properties to *parts* of a thing that only properly apply to the *whole*) and since Bennett and Hacker published the first edition of *Philosophical Foundations of Neuroscience* in 2003 there have been others who have disputed whether the fallacy is a mereological (e.g. John Searle (2007) and Rom Harré (2012) both argue that the problem is not a mereological one because brains are not parts of persons. Hans-Johann Glock also argues that the problem 'cannot be a matter of mereology' because, as Bennett and Hacker grant, psychological predicates *can* be applied to parts of an animal (e.g. in 'Mary's hand hurts') and psychological properties *cannot* be attributed to things like plants and computers. Glock also disputes whether the homunculus fallacy and the mereological fallacy are fallacies at all (see 4–5 of Glock 2020). There is not space in this chapter to get into all of the debates here but my own take is that the criticisms of Searle and Harré do not hit their target. The mereological fallacy, as well as Wittgenstein's remark that inspired it, concern *human beings* rather than *persons* – and a brain *is* a part of a human being (Hacker's response to Harré in 'Before the Mereological Fallacy: A Rejoinder to Rom Harré', *Philosophy*, Vol. 88, Jan 2013 strikes me as a good one). Glock, I think, is on stronger ground, but his own position strikes me as fairly close to that of Bennett and Hacker. Glock agrees with Bennett and Hacker that it is nonsensical to apply psychological predicates to brains and he agrees that it is nonsensical because brains cannot satisfy the (behavioural) criteria for the application of psychological predicates. Bennett and Hacker agree with Glock that the mereological fallacy is not, strictly speaking, a fallacy, although, unlike Glock, they think it leads to fallacies.

41 Boyer 2001, 252.

42 Elsewhere, on 102, he expresses reservations about treating a system in the brain as if it is a human being: '. . . forgive the anthropomorphic tone: the system is of course not aware of what it is doing', he says.

43 Hacker 2018, 249.

44 Thanks to Christopher Hoyt for suggesting that I emphasize *grammar* here.

45 Boyer, P. *Religion Explained*, 98. Justin Barrett makes the same kind of claims in his book *Why Would Anyone Believe in God?*, where he writes that '. . . minds are not accessible to direct investigation and have not even been proven to exist'. (96).

What we see directly, according to Barrett, are (what we take to be) consequences of activities of the mind, that is alterations of physical things, movements of bodies, physical traces of activities and so on. These physical consequences of mental activities 'serve as evidence for their [i.e. minds] existence' (98). Barrett is particularly explicit about the closeness between his own framework and the Cartesian one (although he does not commit himself to Cartesian mind-body dualism). He notes that his own conception of minds and bodies has an ancestor in Cartesian dualism, that we (himself included) find it easy to imagine minds switching bodies, but that we find it difficult ('counterintuitive') to accept that the human mind is embodied (99-100).

46 In Wittgenstein 1993a.

47 Wittgenstein 1993a, 223.

48 See Wittgenstein 2009, §303 '. . . Just try – in a real case – to doubt someone else's fear or pain!'

49 See Wittgenstein 1967a, §225 '"We *see* emotion" – as opposed to what? – We do not see facial contortions and make inferences from them (like a doctor framing a diagnosis) to joy, grief boredom. We describe a face immediately as sad, radiant, bored, even when we are unable to give any other description of the features. – Grief, one would like to say, is personified in the face. This belongs to the concept of emotion.'

50 See Severin Schroeder's book *Wittgenstein* (2006, 208–19) for a discussion of some of the complexities.

51 Hacker 1997, 38.

52 See Wittgenstein 2009 §249 '. . . Lying is a language-game that needs to be learned like any other one' and PPF xi §363, 'A child has much to learn before it can pretend'.

53 See Wittgenstein 2009 §354 and §580.

54 For an excellent detailed discussion of the notion of 'criteria' as it is found in Wittgenstein's work see Lars Hertzberg's (2023) *Wittgenstein on Criteria and Practices.*

55 Wittgenstein 1993a, 298.

4

The brain perceives/infers

Hans Van Eyghen

4.1 Introduction

In a widely influential book,[1] Peter Hacker and Maxwell Bennett argue that cognitive neuroscientists wrongfully attribute psychological states to brains. Robert Vinten applies their arguments to criticize theories in cognitive science of religion (CSR) in this volume. Vinten argues that CSR theorists commit the same mistake when they argue that cognitive mechanisms or inference systems can *infer* or *perceive*. All three authors conclude that the mistake renders a lot of statements in cognitive neuroscience nonsensical. They thereby suggest that the alleged mistake signals a profound problem in cognitive science of religion and contemporary cognitive (neuro)science at large, endangering their validity and use.

Hacker and Bennett's argument has been criticized in a number of ways.[2] I aim to argue that attributing perception and inference to mechanisms in the brain is not problematic. I thereby aim to defend cognitive science of religion and cognitive neuroscience at large. I argue that Vinten and Hacker and Bennett put too many constraints on what counts as a genuine perception or inference. A more liberal, functional concept of both terms allows for a broader, but nonetheless intelligible use.

This chapter is structured as follows: in the next section, I lay out the original argument by Hacker and Bennett; in Section 4.3, I discuss Vinten's updated argument applied to CSR; in Section 4.4, I address Vinten's claim that perception does not involve inferences; in Section 4.5, I discuss what would follow from all arguments if they are sound; in Section 4.6, I criticize the arguments by advancing functional definitions of 'perception' and 'inference'; The last sections (4.7 and 4.8) are devoted to counter objections and replies.

4.2 Hacker and Bennett's argument

Hacker and Bennett argue that the attribution of a host of states to the human brain (or subparts thereof) is thoroughly misguided. They argue that considering the brain as perceiving, inferring, seeing, feeling pain and other mental states is *nonsensical*.[3] In other words, while we know what it means to attribute such states to humans, we really do not when we attribute them to brains. Their arguments draw on older claims by Ludwig Wittgenstein and the broader Wittgensteinian project of analysis of proper use of terms and concepts.[4] Wittgenstein wrote:

> Only of a human being and what resembles (behaves like) a living human being can one say: it has sensations; it sees, is blind; hears, is deaf; is conscious or unconscious.[5]

Since Hacker and Bennett provide distinct arguments against attributing a lot of different states to the brain and discussing all in detail would lead us too far, I focus on their arguments pertaining to perception and inference.[6]

According to Hacker and Bennett the alleged mistake of attributing perception and inference to brains results from the heritage of Cartesianism. Ever since Rene Descartes (1596–1650), it became commonplace to attribute such states to minds and only derivatively to humans. This mistake gradually evolved into attributing such states to brains or subparts of the brain (see the following). Whereas Cartesian dualism between mind and body lost much of its appeal, neuroscientists continued to appeal to a division within the human subject. According to Hacker and Bennett, they did so rather unthinkingly or unreflectively.[7]

As a textbook example of what goes wrong in contemporary neuroscience,[8] Hacker and Bennett cite Francis Crick:

> What you see is not what is really there; it is what your brain believes is there (. . .) Your brain makes the best interpretation it can according to its previous experience and the limited and ambiguous information provided by your eyes (. . .) the brain combines the information provided by the many distinct features of the visual scene (aspects of shape, colour, movement, etc.) and settles on the most plausible interpretation of all these various clues taken together. (. . .) what the brain has to build up is a many-Ievel interpretation of the visual scene . . .. [Filling-in] allows the brain to guess a complete picture from only partial information – a very useful ability.[9]

Hacker and Bennett also refer to a number of other examples.[10] They note a broad consensus concerning the use of terms designating mental states in this way. When applying terms like 'perceiving' or 'inferring' to brains, neuroscientists commit the *mereological fallacy.*[11] Someone commits the mereological fallacy when she ascribes psychological attributes to parts that can only be intelligibly ascribed to a whole.[12] States like perceiving or inferring are the prerogative of human wholes and not that of brains according to Hacker and Bennett.

In a related argument, Anthony Kenny charges contemporary neuroscientists with committing the *homunculus fallacy.* Kenny argues that neuroscientists commit the mistake of ascribing psychological predicates to an imaginary mini-human (homunculus) in the brain. The mistake consists of attempting to explain an action (here perception or inference) by pointing to a subpart that performs the actions.

Doing so just pushes the explanation back a level. For example, explaining perception in humans by arguing that the brain perceives, just begs the question of how or why the brain perceives.[13] Though related, the homunculus fallacy points to a different error than what Hacker and Bennett discuss. Kenny argues that pointing to a subsystem doing *phi* to explain *phi-ing* does not count as a good explanation of *phi-ing* but is intelligible nonetheless. Hacker and Bennett, however, claim that ascribing various psychological states to subsystems is nonsensical.

Hacker and Bennett claim that the term 'mereological fallacy' is more apt than Kenny's term. They add that neuroscientists do not merely commit the mistake of ascribing predicates to parts that are only properly ascribed to wholes but suffer from more general confusion. Their mistake is not merely a form of invalid reasoning but a nonsensical attempt at making a claim. The result is that claims amount to little more than unintelligible nonsense.[14]

Deciding when the use of predicates is apt or not is not an empirical but a philosophical/ conceptual question according to Hacker and Bennett. Just like 'one cannot look for the poles of the earth unless one knows what a pole is',[15] one cannot empirically investigate whether the brain can perceive/ infer without clarity on when the term applies. After all, noting some phenomena that resemble perception or inference might just be examples of mimicking and not real perceptions or inferences.

By calling the mistake of contemporary neuroscientists mereological, Hacker and Bennett imply that psychological states are only properly ascribed to human beings. They argue that while we have a decent idea of what it means to say that a human being is perceiving or inferring, we do not know what we mean when we say the same of brains. We have no good conception of what it means for a brain to process sensory information or weigh arguments.[16] Rather than providing definitions of 'perception' or 'inference' with necessary and sufficient criteria, Hacker and Bennett's arguments rely on a phenomenological description of perception and inference in humans. I turn to their arguments now.

Hacker and Bennett note that we typically come to recognize psychological states in others non-inferentially. We do not come to believe that *S* is in pain, perceiving or inferring because of verbal reports, but usually because we note some outward behaviour. For example, we typically come to believe that someone is in pain because she screams or has a distinctive facial expression and not because *S* states, 'I'm in pain.' Verbal reports or testimony can be regarded as outward behaviour as well.[17] Testimony is, however, not the kind of outward behaviour that directly signals some psychological state. Outward behaviour that points to perception could be the way someone reacts to her environment. For instance, we believe that someone perceives a tree when she does her best to avoid running into it. We also conclude that someone is perceiving a tree when she is focusing her gaze on the tree. Outward behaviour that signals that someone is making an inference would again be by some distinctive facial expression. Hacker and Bennett note that we never see brains that look like they are sunk in thought a la Rodin's *Penseur.*[18] The primary evidence for both psychological states is thus behavioural. Hacker and Bennett argue that we do not usually conclude that a psychological state is present inductively, but non-inferentially and immediately. Beliefs about the presence of psychological states can be overridden by counterevidence.[19]

At this point one might wonder if one cannot conclude to the presence of a psychological state in an inferential way as well. While it might be the case that we often or usually conclude that S is in psychological state x without explicit reflection, it appears that sometimes we do. Sometimes we do know that someone perceived a tree because she said so. We can also draw conclusions about psychological states through more elaborate reasoning. We can rightly conclude that most people who visited the historical city center of Ghent must have perceived the Saint Bavo Cathedral because it is hard to miss. Hacker and Bennett do not deny that inferential attribution of psychological states occurs but state that behavioural components are partly constitutive of what it means to have psychological states. They argue that the criterial (behavioural) grounds for attribution of psychological predicates are *conceptually* connected to those predicates.

This last point points to a tacit view held by Hacker and Bennett where psychological states (like perception and inference) *necessarily* involve bodily behaviour. Hacker and Bennett appear to be arguing that pain without pain behaviour is not really pain. In the same sense, perception or inference without the typical behaviour that accompanies those states in humans is not really perception or inference. As John Searle notes, Hacker and Bennett's view rules out attributing various psychological states to animals with vastly different behaviour than humans, like shellfish.[20]

Most of Hacker and Bennett's arguments concern the attribution of pain or related psychological states. They also have some arguments that specifically target ascribing perceptions or inferences to brains. I discussed how Hacker and Bennett put behavioural constraints on ascribing most psychological states. They do the same for perception and inference. In a specific argument against the possibility of a perceiving brain, they come close to providing a definition of perception. They argue that perception is an epistemic relation between a subject and an object. When a subject perceives a tree, she grasps or understands that tree somehow. Since the object of perception is nowhere to be found inside a human brain, it makes no sense to claim that a brain can perceive.[21]

Hacker and Bennett do not provide a comparable, exact definition of 'inference', but suggest that inference involves forming hypotheses, processing information, reasoning and/or making arguments. They also claim that inferences are transformations of propositions in accordance with a rule. They deny that perception involves any such transformation. Furthermore, they deny that there is anything like hypothesis forming, information processing or argumentation going on inside the human brain.[22]

4.3 Robert Vinten's argument

Hacker and Bennett's arguments are aimed at various theories and approaches in cognitive neuroscience. Their work does not take direct aim at (computational) cognitive science or evolutionary psychology, let alone cognitive science of religion. Robert Vinten applies their lines of reasoning to provide critiques of CSR theories.

I will not discuss CSR theories at length or repeat Vinten's discussion of CSR theories in detail. Instead, I focus on a general feature of CSR theories important for Vinten's argument, that is, that of inference systems. In Hacker and Bennett's grand historical narrative, ascribing psychological states to inference systems can be regarded as the next mereological step. A first mereological step was committed in the early modern era by attributing states to minds rather than humans; a second by neuroscientists by attributing to brains rather than minds; a third by cognitive scientists by attributing to inference systems rather than brains. New approaches thus appear to attribute psychological states to ever-smaller subparts of humans, in this case, subparts of brains.

What are these inference systems?[23] Vinten aptly defines 'inference system' as 'a system in the brain that produces inferences when it is activated by things we perceive in our environment'. The *things* that activate inference systems are triggers. Triggers might be mental objects like reflection or testimony. More often, triggers are observable things in the environment. On one prominent CSR-theory, an inference system is triggered by (vague) patterns or noises.[24] On another, a system is triggered by meaningful events.[25]

When triggered by specific input, inference systems have certain output. According to most CSR theories, that output is (religious) belief,[26] like belief in an invisible agency or God. In Pascal Boyer's theory of intuitive ontological categories[27] (which Vinten discusses at length) the output is a classification of an object in an ontological category.

Apart from being input–output systems, the inference systems discussed in CSR theories have various other properties.[28] (1) One property is that they operate *below the level of consciousness*. The systems produce beliefs or categorizations when properly triggered without the subject knowing why or how. As a result, people can find themselves having certain beliefs or experiences without knowing what triggered them or how they came about.

Another property (2) is that the systems have a *mandatory operation*. The systems do not have a wide range of possible outputs or multiple possible inferences. The route from input to output and the range of possible outputs is rather limited. Inference systems are not mere automata, like puppets on strings that respond to input in one fixed way. They are also unlike humans who can respond to input in a (very) large number of ways, but somewhere in between both.

Inference systems also (3) operate *quickly*. The processing of input and production of output occurs in a (very) short time span. According to defenders of CSR theories, inference systems can produce beliefs when triggered in the blink of an eye.

Inference systems (as discussed in CSR theories) have (4) a distinct evolutionary aetiology. The systems are 'solutions'[29] to evolutionary challenges like the danger of predators or the need for weeding out unreliable cooperation partners. Most CSR theorists argue that inference systems that produce religious beliefs were not selected because of the adaptive value of religious beliefs (if there is any). Religious beliefs are a by-product of those inference systems instead.[30]

According to Vinten, CSR theorists commit the same mistake neuroscientists did before. Pascal Boyer and other theorists use 'familiar expressions in an unfamiliar way'. He argues that it makes no sense to claim that inference systems 'perceive', 'infer' or 'remember'. Such states are only properly ascribed to human beings. Attributing them to subsystems, like inference systems' is committing a grammatical error (see earlier).

Apart from committing grammatical errors, CSR theorists also have a mistaken view of perception according to Vinten. They adhere to a view where perception is closely linked to, or necessarily involving inference. Vinten, influenced by Bennett and Hacker, claims such a view betrays a (mistaken) commitment to a Cartesian model of cognition. I turn to this point in the next section.

4.4 Inference and perception

A close connection between inference and perception expands the scope of (and the importance of) inference greatly. Claiming that all or most perception involves inference at some level is widespread in cognitive science and neuroscience. Hacker and Bennett and Vinten deny it. In this section, I discuss and criticize their argument.[31]

Hacker and Bennett argue that the modern connection between inference and perception draws on seventeenth-century theories of perception. In those theories, sense organs receive certain qualities of objects (primary qualities) which *cause* other ideas in the mind (secondary qualities).[32] Whereas primary qualities are somehow *sent* by objects and picked up by sensory organs, secondary qualities are largely the creation of humans (or their minds). According to early modern theories of perception, humans *construct* an image composed of secondary qualities in a way that is largely unconstrained by the objects themselves.[33] As a result, the image could be profoundly different from the real world.

Hacker and Bennett argue that theories of perception are metaphysical theories and not scientific. They claim that we cannot possibly believe any theory of perception that implies the falsity (or likely falsity) of our beliefs about the world. The normal procedure when we conclude that any belief is false is to discard it. For example, when we learn that we misperceive the lengths of lines in a Muller Lyer illusion, we reject or revise our beliefs accordingly. Nobody (not even scientists who defend theories like the one sketched earlier) discards her beliefs about the external world in a similar way.[34]

Perception is not the effect of an object on us but rather the effect of the actions of objects on our perceptual system according to Hacker and Bennett. They write:

> what we see is not the effect of an object on us. (. . .) Rather *that* we see is a consequence of the action of illuminated or luminous objects on our visual system and *what* we see are those objects, colour and all. What we thus see, we see 'directly' (to see something 'indirectly' might be to see it through a periscope or in a mirror – not to look at the thing itself in full daylight with one's eyes).[35]

Objects do not send data that we can argue from. Instead, we see objects *directly*. We are given perceptions of secondary qualities like redness of geraniums or greenness of grass. These are not caused by humans as a result of contact with objects but are genuine attributes of geraniums and grass.[36]

Hacker and Bennett appear to argue for (or rather state) a form of direct realism and against representationalism. Again, I lack the space to properly assess the arguments for and against direct realism and representationalism.[37] My goal is to argue that representationalism (which is arguably the dominant view in contemporary cognitive [neuro]science) is a viable view. Hacker and Bennett's main beef with representationalism is that representations cannot be ascribed to brains (or subsystems thereof). The way in which defenders of representationalism use the term 'representation' is again misguided and nonsensical. They write:

> [T]he sense in which the excitation of a group of cells represents a certain feature in the visual field is the sense in which a wide ring in a tree represents a year with ample rainfall and that has nothing whatsoever to do with the lexical or semantic sense in which a sentence represents the state of affairs it describes or with the iconic sense in which a picture represents what it depicts.[38]

Scientists are again using terms in a wrongful, ambiguous way when they talk about representations within the human brain. While the brain can contain some form of representing, the brain cannot contain representations of the external world like a picture or a set of sentences does.

In their portrayal of representationalism, Hacker and Bennett do not do justice to how defenders of representationalism use the term. Let us now look at some theories of representation in cognitive science to see how they use the term.

Dan Lloyd argues that the distinctive feature of representations is that they have content about the world.[39] He writes: 'Representation is usually understood as a relationship between something that represents, a "vehicle" (e.g. a picture, sentence, or thought), and what it represents, its "content" (e.g. an object, scene, or situation).'[40] Lloyd adds that the number of vehicles can potentially be unlimited. The role of representations is pinpointing their targets. He explicitly notes that cognitive neuroscientists often employ a functional definition of 'representation', where the identity of a representation and its content depend on its functional role in a broader system.[41]

According to Paul Thagard (computational) cognitive science employs a concept of 'mental representation' akin to how computer scientists use 'data structures'.[42] Like data structures in computers, the mind is storing small chunks of information which can influence belief-formation or behaviour.

Hacker and Bennett appear to have a definition of 'representation' in mind that is too substantial. Their model of representations is a written sentence or a picture, where representations are (putatively) isomorphic to the objects they represent. While these objects represent content, other things (like tree rings) also represent content in a different way. Other forms of representations might be more vague or be further removed from an accurate description of the world. This does not preclude them from representing meaningful information.

Both Thagard and Lloyd define 'representation' by its function rather than by its constituent parts or effects. A functional definition allows for a more liberal attribution of 'representation' beyond what Hacker and Bennett allow. In the following, I will argue that 'perception' and 'inference' can be defined in a similar, functional way. Before I do so, I first discuss the implications of Hacker and Bennett's and Vinten's arguments if they were sound.

4.5 Taking stock

Hacker and Bennett conclude that attributing psychological states like perceiving or inferring to brains is nonsensical. Vinten adds that the same holds for attribution to inference systems as discussed by CSR theorists. Before criticizing their arguments, I first assess what the implications are. I argue that the arguments can imply at least two things. First, they can imply that scientists are making a fairly minor mistake that can easily be rectified by adopting different terms. Second, they can have the far more serious implication that a lot of cognitive neuroscience is radically misguided.

4.5.1 The minor implications

If they accept the soundness of the arguments By Hacker and Bennett and Vinten, cognitive neuroscientists could gracefully adopt new terms. If applying perception and inference to inference systems is indeed nonsensical, all could be made well by using different, sensical terms.

Instead of writing that inference systems perceive, cognitive neuroscientists could start claiming that such systems 'encode stimuli neurally'. Authors like Hacker and Bennett, however, object to the attribution of 'encoding' to inference systems as well. They argue that human subjects do not derive any information from neural representations when they perceive. There is no 'neural code' to be found anywhere in the brain for encoding is a method of encrypting a linguistic expression according to conventional use.[43] Nonetheless, the term 'encoding', is commonly used for artificial processes that try to mimic or copy perceptive systems in computers. Using 'stimulus' and 'neural' is uncontroversial in talk of brain mechanisms.

An alternative for 'inference' could be 'deriving new content'.[44] Non-human animals and systems are commonly said to 'derive' and have 'content', so authors who follow Hacker and Bennett might find use of such terms less objectionable. Still another alternative is suggested by Pascal Boyer. Boyer writes that we need a 'precise description of how brains *receive and organise information*'[45] to understand religious belief.

If the alternatives do not suffice, scientists and philosophers can jointly come up with terms that are acceptable for both. Adopting new, more complex terms will not make the endeavours of cognitive neuroscientists easier and will likely lead to more difficulties in communicating new empirical findings to a broader audience. However, if intelligibility requires it, adopting new terms seems like a fairly easy way out of the problems that Hacker and Bennett and Vinten have with contemporary cognitive neuroscience.

A mere problem with term usage is, however, not what Hacker and Bennett are suggesting and how fellow philosophers have commonly interpreted their arguments. Instead they appear to be suggesting that the wrongful use of terms signals that there is something profoundly wrong with contemporary neuroscience. I turn to this reading of the argument next.

4.5.2 The major implications

When reading Hacker and Bennett's and Vinten's arguments, one cannot escape the sense that they argue for a rather radical revisal or even dismissal of contemporary cognitive neuroscience (including CSR). They write: 'the psychological expressions they use are being invoked in their customary sense, otherwise the neuroscientists *would not draw the inferences from them which they do draw*'.[46] Throughout their discussion, Hacker and Bennett discuss a number of theories. For every theory, they dismiss it for using terms inappropriately, which renders the theory unintelligible or meaningless. After criticizing Pascal Boyer's theory, Vinten asks what the theory really shows if the terms on which it relies are nonsensical.

Although this reading of the arguments is far more damaging than the previous one, it seems more warranted. The use of the terms 'perceiving' and 'inferring' in cognitive neuroscience stretches beyond mere illustration of the mechanisms discussed. Cognitive neuroscientists' use of inference systems 'perceiving' and 'inferring' is also central to their theories. It isn't clear if much can be salvaged of central CSR theories if they have to be reformulated without inference systems with the ability for 'inference'. Most good alternatives will likely suffer from similar objections because their semantic meaning will be rather close to the original terms.

4.6 Reply: Defining perception and inference functionally

Having summarized the charge of committing the mereological fallacy and defended the attribution of representations to human subsystems, I now turn to the central questions of this paper 'Can one properly attribute "perception" and "inference" to inference systems in the brain?' I will argue for an affirmative response to this question that relies on functional definitions. Before doing so, I note that the vast majority of cognitive neuroscientists do not share Hacker and Bennett's idea that attributing 'perception' and 'inference' to brains or subsystems of brains is nonsensical. The wide discussion on cognitivist theories that rely on such attributions is good enough reason to argue that such attribution is sensical. Cognitive neuroscientists appear to have no problem in understanding each other's papers and the theories those lay out. Having a scientific discipline with a growing number of journals, academic centres and scholars would be highly surprising if all were discussing nonsensical theories.[47] The charge of being nonsensical is even more surprising given the fact that cognitive neuroscience also has real-world applications in the form of cognitive behavioural therapy and other therapeutic interventions.

A first prerequisite before one can answer our central questions is when any attribution of a psychological state is proper. My proposal is that attribution of psychological state X is proper when a bearer performs the functions associated with X. For example, something or someone is in fear when it or she displays heightened awareness and restraint caused by a perceived threat or danger. Another example is memory. Someone or something remembers when she (or it) recalls stored information or events. Functional definitions of being in fear or remembering are likely more elaborate. These short definitions, however, suffice to illustrate the approach.

Before we continue, we must note that defining states functionally is not sufficient for all psychological states. A considerable number of psychological states have a distinct phenomenal character that goes beyond their function. A clear example is one of Hacker and Bennett's most discussed states, being in pain.[48] While being in pain has a clear function, that is signalling tissue damage or another (psychological) problem, it also encompasses more. Tissue damage can also be signalled by means of outprints of scans or other forms of medical diagnostics. When one is in pain, one necessarily has a distinct aching feeling or sensation. Without the sensation, pain is not pain.

As a number of philosophers have convincingly argued, the phenomenal character of sensations accompanying psychological states is not required for their respective functions. One argument is known as 'the inverted spectrum'.[49] Psychological states like being in pain or seeing red can be tied to vastly different or inverted conscious experiences. My experience of pain can feel similar to your experience of pleasure while still alerting me to tissue damage. This shows that the phenomenal character of being in pain does not logically follow from or is necessarily connected to its function of signalling tissue damage or discomfort. Another argument for the same conclusion is the 'super-stoic' or 'super Spartan' argument by Hilary Putnam. Putnam evokes a world of super stoics or super Spartans that are able to suppress all behavioural manifestations of pain. They, however, still have the sensation of pain and they dislike it.[50] The possibility of super stoics or super Spartans shows that the function of pain (or rather its effects) can come apart from its phenomenal character.

One might argue that pain consists exclusively of its phenomenal character. Being in pain would merely consist in having the sensation of pain. Clearer examples of composite psychological states with a clear functional role and phenomenal character are available. An example is seeing. Seeing has a functional role, that is, taking in visual information of the environment. Seeing also has a clear phenomenal character. Seeing a geranium as red has a clear first-personal feel about it that is different from seeing greenness or blueness. Reducing seeing to its functional role neglects an indispensable aspect of seeing and is therefore not proper.[51]

Other psychological states (like perception or inference) do not necessarily have a phenomenal character. While perceiving usually has a subjective feel about it for humans (because it is instantiated as visual, auditory, tactile, olfactory or taste perception), perception can merely consist of its functional role, that is, taking in information of the environment.[52] There is substantial empirical evidence for a phenomenon called 'perception without awareness'. In one experiment, stimulus words were presented at such a low level of light intensity that subjects were unable to perceive them at a conscious level. Nonetheless, the stimulus words affected subjects' behaviour. In a second experiment, the number 5 was presented at similar low light intensity. Subjects were nonetheless able to identify the stimulus when their attention was called to it.[53] Other researchers conclude from the data that 'stimuli are perceived even when observers are unaware of the stimuli'.[54]

Since humans appear to be able to take in information from their environment without awareness, perception can be defined merely by its function. A similar argument can be made for inferring. Inferring can be defined functionally as drawing conclusions from or responding to available information. Clear examples of inferences without awareness are reflexes. Reflexes are involuntary, fast reactions to some stimulus. For example, when something is placed in an infant's palm, infants cling to it with their fingers. The infant is usually not aware of the object but its behaviour shows that it inferred that something is there.

As I noted earlier, Hacker and Bennett have a more substantive concept of inference in mind, where inference involves explicit reasoning from premises towards conclusions. Their concept of inference is in line with what professional philosophers are doing when writing academic papers or teaching courses on logic. Usually, humans infer to new conclusions in a far more liberal way, often unreflectively. Defining inference in a limited way like Hacker and Bennett do therefore greatly reduces the scope of inference in a way that seems counterintuitive.

If perception and inference can be defined merely by their respective functions, there is no *a priori* or conceptual reason why perception and inference cannot be properly attributed to brains or inference systems. On many CSR theories, inference systems are able to take in information from their environment.[55] On Barrett's and Guthrie's theory, the agency detection system is able to take in information of patterns or vague noises.[56] On Boyer's theory, the brain is able to take in information of plants, animals or persons and classify it accordingly.[57] None of the inference systems is aware of what or how it is doing so, but otherwise the systems functionally are doing the same, or at least something very similar to what humans are doing when they perceive.

A similar story can be told of inference. The agency detection device concludes from information on patterns or vague noises that an agent is likely around. In Boyer's theory, the brain concludes that something is likely a plant, animal, person or still another thing. Taking on the more liberal concept of inference does not preclude proper attribution of inference to these systems.

The last paragraphs hint at the main obvious reply Hacker and Bennett, Vinten or like-minded thinkers might raise. Why should we accept a more liberal concept of inference or a concept of perception without awareness? I turn to this objection in the next section.

4.7 That is not perception/inference!

I argued in the last section that perceiving and inferring can be properly attributed to inference systems when defined functionally. One might come to Hacker and Bennett's and Vinten's defence by disavowing functional definitions. An objector could argue that *real* perception requires awareness and *real* inference involves explicit reasoning. Functional attribution of both states would be a related state at best. I argue that not only is such a restriction uncalled for, it implies denying Hacker and Bennett's central conclusion that such attribution is nonsensical.

Claiming that functionally defining perception and inference is misguided raises the question 'who has the right to decide on proper definitions of psychological states?' Hacker and Bennett draw on intuitions of a close connection between perception/

inference and their behavioural or awareness components. These intuitions are clearly not shared by a majority of contemporary psychologists and cognitive neuroscientists. Deciding whose intuitions are correct is very difficult. Most contemporary philosophers tend to rely on common sense intuitions when defining their central terms.[58] If we follow this approach, it seems as if the intuitions of empirical scientists are closer to common sense than those of Hacker and Bennett since theirs build on complex argumentation. In any case, the intuitions behind functional definitions appear to have at least some traction, which makes a quick dismissal uncalled for.

More importantly, arguing that perceptions and inferences without awareness or explicit, formal reasoning are not *real* perceptions or inferences acknowledges that such definitions are at least intelligible. We noted earlier that Hacker and Bennett argue that attribution of states like perception and inference to brains is nonsensical. Use of the term 'nonsensical' rose to prominence in philosophical discourse with the logical positivists. They argued that unverifiable sentences are without meaning and hence nonsense. One prime defender of logical positivism, Rudolf Carnap, likened such sentences to the use of the terms 'teavy' and 'toovy'.[59] Both terms lack a clear verifiable meaning and are therefore meaningless. Contemporary objectors will likely not adhere to the verification criterion for which logical positivists were famous. When they argue that functional definitions of perception or inference do not suffice, they tacitly admit that such definitions are at least intelligible. Claiming that functional definitions miss out on an essential feature of perception of inference implies that the definitions are judged as false. Being nonsense or meaningless implies that a proposition cannot be judged as true or false.

Arguing that functional definitions are misguided therefore implies a retreat from Hacker and Bennett's core conclusion that attribution of psychological states to non-humans is nonsensical. It also raises the debate over who has the right to decide on definitions for such states. Both reasons suffice to show that the objection misses its mark.

4.8 The importance of behaviour

Another objection to my counterarguments could state that we should not attribute psychological states to inference systems (and brains) because they do not display the behaviour that commonly accompanies those states. Hacker and Bennett discuss at length how states of pain are usually accompanied by behaviour like crying, twitching or refraining from action[60] (Bennett and Hacker 2003: 82). Since inference systems do not display such behaviour, attribution of psychological states to them is unwarranted.

Vinten suggests a similar reply when he notes that inference systems do not exhibit the kind of behaviour that living humans do. Like Hacker, Vinten notes that (unlike brains and inference systems) human groups can and do exhibit behaviour which is usually associated with mental states like 'perceiving', 'remembering' and 'desiring'. It makes sense to state that 'the schoolchildren *saw* the dolphin leap out of the water' because they collectively display facial twitches or other behaviour that matches perception. Groups can also display behaviour commonly associated with 'inferring'.[61]

Vinten adds that we would not *know* what behaviour would be involved when a brain is in a similar state or what it would be like for a brain.[62]

By stating this objection, the objector would move from a conceptual discussion over what concepts like 'perception' or 'inference' mean to an epistemological claim. Claiming that attribution of states to subsystems is unwarranted because they lack the behaviour that indicates such states boils down to a claim that we lack sufficient evidence for attribution. If we can only properly attribute perception to beings that display certain behaviour, that behaviour is regarded as *evidence for* a psychological state. Absence of such evidence would then preclude attribution.

Again, this objection moves into a rather different dialectic than was apparent in Hacker and Bennett's original argument. Rather than claiming that attribution is *meaningless* because psychological states are conceptually connected to behavioural components, we moved to a claim that attribution is *not warranted* if the requisite behavioural evidence is missing.

Inference systems do not display any behaviour apart from their internal operations. While their operations could correspond to neural changes, they do not display similar behaviour which humans display when they perceive or infer. Contrary to Hacker and Bennett, many philosophers and cognitive scientists do not share the intuitions that perception or inference needs to involve bodily behaviour. I noted earlier how Putnam raises the possibility of super stoics or super Spartans. While they have the same pain sensations ordinary humans have, they learned to suppress their bodily reactions.[63] Putnam thereby suggests that pain behaviour is not quite requisite for being in pain.

Turning the discussion back to perception and inference, we need not imagine a special kind of humans. We have plenty of examples of humans making inferences in their heads without any noticeable outward behaviour. People make quick calculations all the time without any noticeable face twitches or movements. They also routinely infer to new information when reading or watching the news without changes in their facial expressions or other bodily actions. The same holds for perception. Humans perceive objects and persons all the time in waking state. While this often impacts their behaviour (e.g. by avoiding collisions), equally often it does not. Sometimes people just perceive things and continue their affairs without any behavioural change. In many of these cases, it remains obscure to outsiders if subjects were perceiving or inferring, let alone what they were perceiving or inferring to. Regardless we can *know* that subjects were in such states by generalizing from our own experience or by testimony thereof.

I believe the examples also suffice to show that an even stronger case can be made that perception and inference need not involve bodily behaviour than can be made for other psychological states like being in pain. In any case, the claims that attribution of inference or perception without bodily changes is unwarranted do not hold water.

Vinten argues that behaviour can be regarded as *defeasible* evidence for the presence of a psychological state. In some cases, humans can be deceitful or disguise their inner life. According to Vinten this does not preclude logical relationships between behaviour and mental going-ons. Vinten could argue that 'perceiving' and 'inferring' are necessarily joined to particular outward behaviour. In cases where such behaviour is seemingly absent (see earlier), the behaviour needs to be overridden or forcefully suppressed. When humans thus perceive or infer without any noticeable behaviour the normal course of action (which does involve behaviour) gets distorted or corrupted. While consistent, this account is not plausible. There is no strong evidence that humans who perceive or infer without behaviour are suppressing anything. Given the commonality of perception and inference it is more likely that most perceptions or inferences occur *without* any bodily behaviour.

4.9 Sub-personal explanation versus personal attribution

Hacker and Bennett's original argument drew criticism from a number of famous philosophers. One prominent opponent, Daniel Dennett, makes several points that are akin to my counterarguments. However, a close reading of Dennett's argument shows that he is in fact agreeing with Hacker and Bennett's conclusion that ascribing psychological states to brains is misguided. According to Dennett, use of psychological terms is only appropriate *analogically* in sub-personal explanations of these states.

Dennett agrees that humans may feel pain or see things but their brains do not. He argues that there is no event or process in the brain that exhibits the characteristics of these states.[64] So far, Dennett appears to be in agreement with Hacker and Bennett. He adds though that one can note phenomena or behaviour that *resembles* that of humans. According to Dennett, the resemblances are sufficient to warrant an adjusted use of psychological vocabulary applied to brains and their subparts.[65]

Use of (adjusted) psychological vocabulary is admissible in sub-personal explanations of phenomena or behaviour observed at the personal level. Sub-personal explanations are explanations that focus on the level of brains and events in the nervous system.[66] For example, a sub-personal explanation of visual perception consists of the operations of the human eye and the mechanisms in the visual cortex.

Dennett appears to allow attribution of psychological states to (sub-personal) human subsystems. His motivation for doing so is not because they fit necessary and sufficient criteria, but because they resemble states in (personal) humans. Dennett's remarks that psychological terms should be used in an *adjusted* way. I agree that this is the case for psychological states with an indispensable phenomenal character, like being in pain or hearing. However, my account allows that states without an essential phenomenal character can be attributed to subsystems *univocally*.[67] Dennett's insistence on an adjusted use of terms suggests that he, at best, allows for an *analogical* use.

The main difference between Dennett's arguments and mine is that mine allows for full attribution of states like perceiving and inferring when it meets the functional criteria of these states. Attribution is not made proper on the mere basis of resemblance but on a basis of definition. Defining perceiving and inferring functionally allows one to distil necessary and sufficient criteria for perceiving and inferring. These can serve to assess whether any inference system can perceive and/ or infer.

4.10 Rephrasing definitions leads to confusion

In older work, Hacker seems to object to a change in definitions of psychological states. Diverging from common-sensical use of terms would lead to conflicts of use or incoherent use of terms. I argue that stating definition more clearly, if anything, reduces confusion and incoherent use of terms.

Several years before his collaborative work with Bennett, Hacker wrote:

> If neurophysiologists, psychologists, artificial-intelligence scientists, or philosophers wish to change existing grammar, to introduce new ways of speaking, they may do so; but their new stipulations must be explained and conditions of application laid down. What may not be done is to argue that since we know what 'to think,' 'to see,' or 'to infer' mean and know what 'the brain' means, therefore we must know what 'the brain thinks, sees, and infers' means. For we know what these verbs mean only in so far as we have mastered their existing use, which does not license applying them to the body or its parts, save derivatively. Nor may one cross the new 'technical' use with the old one, as, for example, neuroscientists typically do in their theorizing. For this produces a conflict of rules and hence incoherence in the neuroscientists' use of these terms.[68]

In this paragraph, Hacker argues that (1) scholars may change how terms are used if they lay down the conditions of application. He also suggests that (2) the meaning of verbs must not be changed in a way that runs against accepting existing use. Both claims are compatible, but (2) puts serious constraints on (1).

As to when terms are redefined in such a way that runs against common sense, vagueness arises. Dennett rightly argues that revision of terms often occurs because of new empirical discoveries.[69] Definitions of 'perception' also changed from the way it was used by ancient or medieval philosophers. If we follow Hacker, change of definitions of psychological states is hardly ever warranted. Not only are Hacker's intuitions not widely shared but his claims also close the door to conceptual change due to new discoveries or changes in philosophical outlook.

Apart from these worries, I argue that defining perceiving and inferring functionally adds to clarity and thereby avoids confusion. With clearer definitions, we can better assess when an organism or a subsystem perceives or infers. To do so, we need not rely on a shared common sense concept, which is often vague and hard to spell out.

4.11 – Conclusion

I argued that attribution of psychological states like 'perceiving' and 'inferring' to brains or inference systems in the brain is appropriate if the terms are defined functionally. A functional definition denies the intrinsic connection of both states to bodily behaviour. As a result, claims like 'the brain perceives' or 'the brain infers' can be made sensically without the risk of confusion.

Notes

1 At the time of writing, Google scholar counted 2798 citations of Hacker and Bennett's book.
2 Rom Harré argues that brains can be regarded as tools by means of which human persons argue, infer or perceive (Harré 2012). Daniel Dennett argues that states of brains resemble those of persons at a lower level of analysis (see the following) (Dennett 2007). John Searle argues that Hacker and Bennett's argument prevents attribution of various states to non-human animals (Searle 2007). Figdor argues that Hacker and Bennett's do not do justice to accumulated knowledge on various states and their proper domains (Figdor 2018). Keestra and Cowley argue that psychological states can be usefully attributed as heuristics in cognitive neuroscience (Keestra and Cowley 2011). Hacker and Bennett reply to a number of these objections in the second edition of their book (Bennett and Hacker 2022).
3 The summary of Hacker and Bennett's arguments relies mostly on their 2003 rendition. Where needed, some elements from the second edition (2022) were included. Given that the arguments are largely the same in both editions, my counterarguments apply to the second edition as well.
4 Hacker and Bennett note that their claim has ancient roots in the works of Aristotle. Aristotle argues that it is a mistake to attribute states like 'pitying', 'learning', and 'thinking' to anything else than human beings. Attributing to a human subsystem like the 'psyche' is a mistake (Bennett and Hacker 2022).
5 Wittgenstein 1958, §281, quoted by Bennett and Hacker 2022, 82.
6 All arguments appear to share a common structure where various attributions of states are shown not to fit the phenomenological requirements for attribution of such states. The phenomenological requirements vary considerably.
7 Bennett and Hacker 2003.
8 Hacker and Bennett note that the mistake is also common amongst psychologists (Bennett and Hacker 2003, 73 footnote 14.) Their arguments can also be applied to cognitive scientists as done by Robert Vinten (see the following).
9 Crick 1995 quoted by Bennett and Hacker 2003, 68.
10 See (Bennett and Hacker 2003, 68–9).
11 Bennett and Hacker 2003, chap. 3, 68–107.
12 Smit and Hacker 2014, 1077.
13 Kenny 1971.
14 Bennett and Hacker 2003.
15 Bennett and Hacker 2003, 71.
16 Bennett and Hacker 2003, 179.
17 Hacker and Bennett acknowledge that verbal behaviour can be a criterion for attributing mentality to others (Bennett and Hacker 2003). They, however, suggest that this is not how humans typically attribute psychological states to others.
18 Bennett and Hacker 2022, 12, 97, 189–90.
19 Bennett and Hacker 2003, 82.
20 Searle 2007, 104.
21 Bennett and Hacker 2003, 128.
22 Bennett and Hacker 2003, 135–7.
23 Inference systems are more commonly called 'cognitive mechanisms' in CSR-literature. In the remainder of this paper, I go along with Vinten's and Boyer's use of the term 'inference systems'.

24 Defenders of the theory argue that such triggers trigger that *Hyperactive Agency Detection Device* (Barrett 2004; Guthrie 1993). Triggers like these are read as signs indicating the presence of agency. When corroborating evidence of agents is absent, the system may foster or give rise to beliefs in invisible agency, like that of spirits or gods (Guthrie 1993; Barrett 2004).
25 Jesse Bering argues that meaningful events may trigger a system to (unconsciously) infer to an ultimate meaning-giver (Bering 2002).
26 Some theories argue that the output is religious or ritual behaviour (e.g.: Boyer and Liénard 2006).
27 Boyer 2002.
28 The discussion on properties of inference systems is drawn from (McCauley 2011). McCauley also discusses other properties, namely domain specificity, informational encapsulation, shallow outputs, fixed neural architecture, characteristic patterns of breakdown and characteristic development. Most of these are in turn drawn from Jerry Fodor's original account of modular mechanisms (cf. Fodor 1983). I do not discuss these because they are not important for my following arguments.
29 Properly speaking, natural selection does not solve evolutionary problems but retains traits that help organisms increase fitness. Samir Okasha argues that anthropomorphic terms are sometimes inevitable in evolutionary discourse (Okasha 2018). Ascribing goals or agency to evolution probably does not stretch beyond metaphorical use.
30 Boyer 2003.
31 Hacker and Bennett also discuss (the attribution of) sensations at length. Unsurprisingly, they argue that sensations are only properly attributed to human beings. I do not discuss this argument nor do my criticisms apply to it.
32 Bennett and Hacker 2003, 128–35.
33 I lack the expertise to properly assess whether the portrayal of seventeenth-century theories of perception is correct.
34 Bennett and Hacker 2003, 131.
35 Bennett and Hacker 2003 on page 131 of Bennett and Hacker's *Philosophical Foundations of Neuroscience* (2003) they say "*What* we see is not the *effect* of an object on us. The effect of an object on our nervous system is the stimulation of the cells of the retina, the effect of this on the optic nerve, the consequent excitation of the cells in the hypercolumns of the 'visual' striate cortex – but none of this is perceived either by the brain (which can perceive nothing) or by the person whose brain it is. Rather, *that* we see is a consequence of the action of illuminated or luminous objects on our visual system, and *what* we see are those objects, colour and all. What we thus see, we see 'directly' (to see something 'indirectly' might be to see it through a periscope or in a mirror – not to look at the thing itself in full daylight with one's eyes)."
36 Hacker and Bennett write: the red colour of the geranium does not have to be *transmitted* to me in order for me to see it. Nor does it have to be *caused* in me. For the red I see when I see the geranium is neither *in* me nor *in* the geranium. It is not *in* anything. But it is an attribute *of* the geranium, not of my perception of the geranium (2003, 132).
37 For an overview, see: (Craig and Crane 2016).
38 Bennett and Hacker 2003, 142.
39 Hacker and Bennett echo this view when they suggest that representations should 'stand for something'. They write: 'The firing of cells in V1 may be signs of a figure with certain line orientations in the animal' s visual field, but they do not stand for

anything, they are not symbols, and they do not describe anything.' (Bennett and Hacker 2021, 88).
40 Lloyd 2006, 1.
41 Lloyd 2006.
42 Thagard 2012.
43 Bennett and Hacker 2022, 180.
44 An alternative could be 'building new representations'. However, Hacker and Bennett also deny that attributing representations to human subsystems is intelligible (2003). Elsewhere, Hacker and Smit note that Daniel Dennett and John Searle propose to regard the brain as an '... organ that receives information and uses processed information for planning action' (Smit and Hacker 2014, 1089). Hacker and Smit response that 'information processing' is also meaningless when attributed to brains.
45 Boyer 2001, 3.
46 Bennett and Hacker 2022, 87.
47 A similar point is suggested by Carrie Figdor. Figdor notes how Hacker and Bennett's arguments would imply that the works of many Nobel prize winners are nonsensical (Figdor 2018).
48 Bennett and Hacker 2003.
49 The argument was first raised by John Locke (Locke 1979). For an overview, see (Byrne 2004).
50 Putnam 1963.
51 Hacker and Bennett object to talk about the phenomenal character of experiences or 'qualia'. They argue that experience cannot be defined or identified by the affective emotions that come along with them. Questions regarding what it is like to have an experience would sound strange to common people and would resist clear answers (Bennett and Hacker 2022, 301-21). Instead experiences ought to be specified 'by what they are and not how they feel'(Bennett and Hacker 2022, 311). A thorough discussion of their argument lies beyond the scope of this chapter. I merely note that philosophers who discuss qualia usually do not argue that qualitative character is all there is to an experience and hence can be defined by it. They merely argue that subjective experiences have a particular subjective sense about them. This subjective sense is different than an objective third-person description of something. Whether this subjective sense can be clearly articulated is unimportant.
52 Defining perception as taking in information of one's environment excludes taking in information of one's own internal life. This form of perception is more commonly called 'introspection' and is very different from ordinary perception.
53 Banreti-Fuchs 1967.
54 Merikle, Smilek and Eastwood 2001.
55 On a functional definition of 'perception', perception need not involve having perceptual organs like eyes, ears or skin. It suffices that perceiving objects like inference systems receive information. For example, the information could be transmitted via neurons.
56 Guthrie 1993, Barrett 2004.
57 Boyer 2002.
58 See (Peels and van Woudenberg 2020) for various examples.
59 Carnap 2021.
60 Bennett and Hacker 2003, 82.
61 One can think of Monty Python's 'Philosophy Football' sketch to get an idea of the behaviour of group inference.

62 Most of Vinten's examples pertain to 'responding', 'speaking' and 'perceiving'. His line of reasoning can be applied to 'inferring' as well.
63 Putnam 1963.
64 See: (Dennett 2010).
65 Dennett 2007.
66 See Drayson 2014.
67 Figdor concludes that attributions can be read literally as well (Figdor 2018).
68 Peter Hacker *Wittgenstein: Meaning and Mind* 1990, 60–1 cited on p. 85 of *Neuroscience & Philosophy* in Dennett 2007. The original edition of Hacker's book was published in 1990. A new revised edition was published in 2019 (the corresponding passage there is on pages 90–1).
69 Dennett 2007.

5

The imaginary inner inside the cognitive science of religion

Christopher Hoyt

5.1 Introduction

As Justin Barrett notes, the cognitive science of religion (CSR) appears to be converging on the thesis that because of our evolved faculties of mind, human beings are predisposed to believe in gods and other supernatural phenomena.[1] Most CSR theorists fall into one of two camps: those who hold that religious beliefs are a by-product of cognitive systems evolved for other purposes, and those who believe that religious beliefs and practices themselves serve some adaptive need. But virtually all CSR theorists agree that religious belief emerges out of mental faculties that evolved long ago in human history and that the central project of CSR is to describe those faculties and the processes by which they produced religious belief and practice. From a Wittgensteinian point of view, CSR appears to be deeply misguided and confused. The causal explanations of religious belief promised by CSR are not really suited to answer the questions at hand, and the mental faculties posited by CSR are ontologically unfit to do the work claimed of them. To understand religious belief and practice, Wittgenstein's philosophy suggests, we are better off employing entirely different modes of explanation. Real understanding of religious beliefs mainly comes out of perspicuous representations of religious phenomena, not instrumental accounts of their underlying causes.

Why are Wittgenstein and CSR so far apart? The question is a good one since both Wittgenstein and CSR seem to begin with roughly the same naturalistic view of religion. In his 'Remarks on Frazier's *Golden Bough*', Wittgenstein consistently urges us to see that religion is an 'extension of instinct' and that religious life and language evolved through a natural-historical process.[2] It is the combination of our animal nature and our historical past that explains the formation of particular religious phenomena, Wittgenstein suggests. Religious ritual, says Wittgenstein, springs from the awe and reverence naturally inspired by fire, sex, birth, death and the many other phenomena important in human life around which rituals are built. Rituals might be called 'instinct-actions', he argues, comparable to beating the ground in spontaneous expression of anger.[3] Wittgenstein writes:

> [The] characteristic feature of the awakening mind of man is precisely the fact that a phenomenon comes to have meaning for him. One could almost say that man is a ceremonial animal. . . . That is, one could begin a book on anthropology by saying: When one examines the life and behavior of mankind throughout the world, one sees that, except for what might be called animal activities, such as ingestion, etc., etc., etc., men also perform actions which bear a characteristic peculiar to themselves, and these could be called ritualistic actions.[4]

Our ritual responses rise up naturally in response to our context. A forest-dwelling people that is 'united in a community of life' with its trees might worship the oak, for example, and 'If fleas developed a rite', Wittgenstein continues, 'it would be based on the dog'.[5] However, there is no room in Wittgenstein's philosophy for the sort of psychological explanation of religion that CSR theorists offer. To understand why, let us focus on the work of one particular CSR theorist, Pascal Boyer. Given our general interest, Boyer's work is suitably representative while the relevance of Wittgenstein to CSR will only become apparent when we look at specific problems and theories.

5.2 Perspective posing as science

Boyer is a CSR theorist of the first variety, viz. those who believe that religious beliefs are the byproducts of cognitive systems that evolved for other purposes.[6] Just as the brain evolved to support language and we got music as a happy side effect,[7] Boyer maintains that sundry cognitive systems developed to yield a wide variety of evolutionary advantages and only accidentally gave us religion as well. For example, our anxiety about corpses probably has its roots in a fear of contagion, but it is also the foundation of burial rituals.[8] To take another example, consider that evolution provided us with a host of moral feelings to support social cooperation – guilt felt for cheating others, pride for helping and so on. According to Boyer, one function of the concept of *god* is to justify or substantiate our intuitive moral feelings, to offer us a way of rationalizing those feelings.[9] In this and many other cases, our religious ideas and practices are parasitic upon prior cognitive systems.

Not all CSR is hostile to religion, but a lot of it is, and Boyer's condemnatory view is largely representative of the field. Boyer regards religious beliefs and practices as the effects of mental blunders, of cognitive misfires. For example, Boyer argues that gods are sometimes introduced into rituals to satisfy an instinctive but misguided yearning to posit agency and causality where they don't actually exist. Thus, people are prone to the false belief that their ritual actions can have causal effects in the world via the mediate agency of the gods.[10] Why are people prone to believe such obvious nonsense? Boyer blames our evolved *inference systems*. These are the supposed mechanisms of mind that produce our thoughts and feelings. According to Boyer, there are two overactive inference systems to blame for the false idea that rituals can invoke divine causes. First, the *causal inference system* inclines us to see causal connections where none exist. Rituals look like things that have effects, and due to the burden of an impetuous *causal inference system*, we are apt to see cause–effect relationships that simply are not there. Second, when faced with significant events such as illness or death that we cannot otherwise explain, we are likely to assume that a malicious actor must be to blame. Boyer writes:

> What matters to rituals and makes them relevant is that one construes the social effects as the *result* of the actions prescribed. This inevitably creates a causal gap. Because of the massive salience of agency in our mental systems, most humans fill this gap with concepts of agents.[11]

In other words, the belief that gods are the causal agents at work in ritual amplifies the more basic mistake of thinking that rituals have any causal power at all. While science offers us a way to overcome such bungles and see the world aright, religious believers are the epistemological victims of their animal instincts.

The grand picture behind all of Boyer's work is the idea that the 'mind is a complex set of biological machines that produce all sorts of thoughts. For many thoughts there is no reasonable reason, as it were, except that they are inevitable results of the way the machines work'.[12] The machinery that Boyer imagines operates below the limits of consciousness in the 'mental basement'[13] where inference systems continuously calculate, scheme, judge and perform other cognitive functions but only reveal themselves to our conscious minds through their products. In social contexts, for example, your unconscious mind continuously performs 'precise calculations' about the intentions and trustworthiness of others, according to Boyer. Yet you consciously experience only 'vague concepts' like 'this person is "likeable," [and] this group is "friendly"'. Here and elsewhere, we suffer a 'lack of access to inference systems' that guide our actions, and instead consciously experience only emotions and intuitions.[14] Boyer's primary aim is to lay bare the precise workings of these unconscious inference systems by making use of the 'experimental resources of cognitive science'.[15]

Boyer's reductive and dismissive account of religion is rooted in the particular inference systems that he posits as the causes of belief. Take, for example, Boyer's account of burial and funerary rites. These are rooted, says Boyer, in the confused thoughts and feelings we have when we see corpses, especially those of loved ones. Our reactions are explained, says Boyer, by the inconsistent behaviour of at least three distinct inference systems:

(1) The *intuitive psychology system*, which 'creates automatically a particular description of what a situation is like as seen by the people we interact with'.
(2) The *animacy system*, which 'is activated by the sight of any object that moves in a purposeful manner. It produces expectations and inferences about animals and persons'.
(3) The *person-file* system, 'a kind of mental Rolodex or *Who's Who* of the person's social environment'.[16]

When we look upon the corpse of a beloved, says Boyer, our complicated reactions are explained by the inconsistent outputs of these three systems. The animacy system is 'quite clear' that the dead are 'ex-persons' without goals, etc. On the other hand, the 'person-file system just cannot "shut off"'. It keeps producing inferences about the deceased as though he were still alive. 'A symptom of this incoherence is the hackneyed phrase we have all heard or used at funerals: "He would have liked it this way."'[17] We are inclined to respect the wishes of the dead, then, because our inference systems are inadequately evolved to handle the sight of a corpse. The *person-file system* is clever in how it deals with the living, but it can't keep up with the *animacy system* when it comes to dealing with the dead. According to Boyer, the religious belief that the soul can survive the death of the body derives from this same confused, inconsistent pattern of inferences going on in our unconscious minds. The 'confused impression' that the corpse maintains a shade of life is the by-product of the unfortunate confluence of our 'incompatible intuitions about dead persons'.[18]

Boyer's comment that you are *confused* if you say about a deceased loved one that 'he would have wanted it that way' is strange. The expression, when used sincerely, is so clearly one of piety that it's hard to stomach or to understand Boyer's insensitivity. I don't mean merely that Boyer has been impolite, but that he seems blind to the actual meaning of the expression. If man is a ceremonial animal whose mind is awake to meanings, Boyer's response sounds pretty sleepy. A grieving atheist wanting to express his love and his sorrow might explain his choice to serve drinks at his father's wake by saying, 'he would have wanted it that way'. The son need not be confused, there need not be even a small part of him that believes his father will judge or punish him. The words are an expression of sorrow and reverence that have little or nothing to do with any specific beliefs about the causal or cognitive powers of the dead. The meaning of the expression might be better explained by comparing it to the idiosyncratic gesture of respect made by Schubert's brother upon his death. Wittgenstein writes:

> Recall that after Schubert's death his brother cut some of Schubert's scores into small pieces and gave such pieces, consisting of a few bars, to his favorite pupils. This act, as a sign of piety, is just as understandable to us as the different one of keeping the scores untouched, accessible to no one. And if Schubert's brother had burned the scores that too would be understandable as a sign of piety.[19]

The brother's gesture was one of respect and grief, not misguided magic. It was an attempt to express his love and his pain and to locate his brother's death within the context of his own life, in a manner of speaking. 'Someone who is affected by the majesty of death can give expression to this through [his manner of] . . . life', Wittgenstein remarks a little earlier in the same text.[20] The concept of a mistake has no foothold here.[21] The same thing can be said about many formal religious ceremonies. Many Jews who don't believe in an afterlife still perform funerary rites and pray for the dead.

Boyer's theory and his blindness to the religious meanings are both strongly reminiscent of James G. Frazer, who posits not three *inference systems* but two principles of magical thinking, two principles that supposedly explain the specious causal reasoning that underlies 'primitive' religion and ritual. First, the Law of Similarity states that an action or an object tends to produce its likeness. Second, the Law of Contagion states that things once in contact tend to continue to effect each other through action at a distance. Based on these two laws, Frazer claims to explain why so many 'primitive' peoples believe that they can causally control the rain, for example: Russian peasants try to conjure rain by drumming out the sound of thunder; Omaha Indians spill a barrel of water near their crops in the hope of drawing water to water; Zulus evoke the tears of heaven by sacrificing a sacred bird.[22] Wittgenstein chastises Frazer for missing the real meaning of these rituals and for making them look stupid through his own misunderstanding. Were those 'primitives' so foolish as Frazer suggests, says Wittgenstein, surely they would perform their rain dance at the beginning of the dry season and not only after it has come to cause real suffering.[23]

Whereas Frazer claims to have discovered the underlying psychological mechanisms that render 'natives' susceptible to foolish habits of mind, Wittgenstein says that what he has actually done is to lay out a certain perspective, a way of seeing the facts at hand, but it is just one way amongst many. 'The historical explanation, the explanation as an hypothesis of development,' says Wittgenstein 'is only *one* way of assembling the data – of their synopsis'.[24] There are many ways of looking at a rain dance or at a man saying of his dead father 'he would have wanted it that way', but Frazer and Boyer insist that such occasions be regarded from just one condescending vantage from which they appear irrational. Over and over, the two disregard the perspectives that make sense of religious action in favour of a perspective that makes them look dumb. In a long passage that is particularly helpful here, Wittgenstein writes:

> The concept of a perspicuous representation is of fundamental importance for us. It denotes the form of our representation, the way we look at things. . . . This perspicuous representation brings about the understanding which consists precisely in the fact that we 'see connections'. . .
>
> But an hypothetical connecting link should in this case do nothing but direct the attention to the similarity, the relatedness, of *facts*. As one might illustrate an internal relation of a circle to an ellipse by gradually converting an ellipse into a circle, *but not in order to assert that a certain ellipse actually, historically, had originated from a circle* (evolutionary hypothesis), but only in order to sharpen our eye for a formal connection.
>
> But I can also see the evolutionary hypothesis as nothing more, as the clothing of a formal connection.[25]

In other words, the evolutionary hypothesis might take on the false appearance of natural history when its real function is to yield a particular outlook, a vantage from which to regard religious forms of life.

Boyer's proposed faculties of mind are really nothing more than the clothing of a certain outlook that is numb to religious meanings from the outset. Each of the *inference systems* that he posits is really only a metaphor in the guise of an evolutionary hypothesis. People often behave *as though* they were compelled to see things from the perspective of their fellows; people often respond to others and to objects *as though* they could not resist positing animacy; people often behave *as though* they carried a Rolodex around inside them. These are metaphors that Boyer parades as scientific discoveries. The real value of a metaphor is to guide us to a particular way of seeing matters, and Boyer constantly applies his in the service of showing religious belief and action in the most unflattering light. Boyer encourages us to see belief in God as an exact parallel to belief in Santa Claus, for example, but he dresses his perspective in the formal clothing of an evolutionary-psychological hypothesis about the supposed cerebral processes of ontological thinking that lead to misfires that spit out beliefs in God and Santa at different moments.[26] He invites us to see belief in a judicial god as the hypostatization of our moral intuitions.[27] He invites us to see rituals as meaningless analogues to obsessive-compulsive behaviour (OCD) by postulating a single cerebral dysfunction at the core of both.[28] In these and many more cases, Boyer offers us these metaphors dressed up as evolutionary theory.

It is important to acknowledge that if we regard Boyer's *inference systems* as metaphors for tendencies in human thought and behaviour, as we should, they can be helpful. Boyer is certainly right that our evolved reactions of disgust to maggots, rotting meat and so on are related to our burial rites. It is nearly impossible to imagine a people who would simply leave their dead to decompose wherever they might fall, and Boyer is right to suggest that our common reactions are rooted in instinct. However, there are miles of open road between the fact that there we are instinctively disgusted by the smell of rotting flesh and the claim that such disgust will inevitably feed into a process of inference that will lead a person (at least a layman untrained in science) to a state of cognitive dissonance when a parent dies. When we look at that hypothesis as the dressed-up metaphor that it is, we see that its real effect is to render us blind to the aspect of the funerary rites he claims to explain. If Boyer can't follow why someone would say at a loved one's funeral that 'he would have wanted it that way', that is not because he has transcended his 'primitive' instincts through the power of science but because he has lost touch with certain dimensions of human experience. In this regard, Boyer makes exactly the same mistake that Wittgenstein attributes to Frazer: 'Frazer is much more savage than most of his savages [whose rituals he describes]', writes Wittgenstein, 'for they are not as far removed from the understanding of a spiritual matter as a twentieth-century Englishman'.[29] Contrary to Frazer and Boyer, Wittgenstein consistently aims to help awaken our modern sensibility to the wondrous aspects of the world that serve as the grounds of religion. Why do 'primitives' burn effigies of their enemies? Their actions are not confused attempts at magic, Wittgenstein remarks, but rather more like kissing the photograph of one's distant beloved.[30] A primitive rain dance might be no more based on false causal thinking than singing hymns in church is meant to make a congregation wealthy.

In the next two sections, I aim to clarify and deepen the points introduced so far. First, we turn to evidence from evolutionary psychology itself that Boyer's theories are conceptually confused. Briefly put, Boyer borrows terms from ordinary language that evolved to operate in social contexts and he takes them on holiday assuming from the outset that they refer to theoretical posits. Then, in the final section of the chapter, I describe in more detail and with greater nuance the nature and form of understanding the religious beliefs and practices of other people.

5.3 The snake eats its own tail

To explain the fundamental error of Boyer's psychological hypotheses I draw from insights made by evolutionary psychology itself regarding our evolved capacity to see and to talk about other minds. This strategy is intended to do double-duty: first, it elucidates why it is incoherent to construct a theory of hidden mental mechanisms using a vocabulary of ordinary language psychological terms like 'thinking' and 'calculating'. Second, it demonstrates how evolutionary psychology can be done well and provide valid insights relevant to a wide variety of subjects, including religion. Furthermore, in combination these two insights allow me to give credit where credit is due. Boyer's research can and does offer some valid insights into religious life that remain after we subtract his untenable hypotheses and explanations.

I want to borrow and expand on an idea proposed by John Canfield in a kindred attempt to triangulate Wittgenstein, evolutionary psychology and religion. Canfield meshes a speculative account of early hominid language development with current primate research to explain how pronouns like *I*, *me* and *you* likely evolved. Pondering the proto-languages of early hominids, Canfield asks how our distant ancestors might have first begun to graft systematic vocalizations onto existing patterns of social interaction. Following Wittgenstein, Canfield proposes the following model: 'Both the before-language hominid and the pre-linguistic child engage naturally in certain behavior patterns basic to speech.'[31] Instinctively, vocal gestures are integrated into these behaviour patterns. Whines, grunts and cries are basic to human life and are elicited by reasonably constant stimulus conditions. Gradually, those vocalizations become socially constrained and eventually they evolve into words. 'In the simple language-game, the symbol qua symbol takes over the role of gesture.'[32] Mother and infant exchange gestures of pleading and offering which include smiles and wide eyes, whines and coos. Indeed, primate research shows that chimpanzee mothers and infants engage in behaviour patterns, including vocal behaviours, much like those of humans.[33] Gradually, however, the human babies are trained into patterns that the chimpanzee babies are unable to acquire, especially increasingly specific and complex patterns of vocal behaviour.[34]

The foundations of language, Canfield proposes, likely lie in the basic behavioural patterns acquired early in life: requesting, greeting, naming, refusing, pretending and possessing.[35] At much later stages of development, language allows us to describe ourselves and others in longer narratives that extend beyond the present in both time and space while seeming to track a single *self*. The *I* of my childhood is spoken about as though the same *I* exists within me now in a transformed body and context. Through such linguistic patterns, the myth of the inner *self* emerges with such seeming certainty that it is almost irresistible.[36] If true, Canfield's account might help make sense of why the urge to misunderstand the logic of our language is so powerful that it seems to have a hold over people from every culture. As we will see later, the CSR theorist Justin Barrett walks right up to the line, seeing that the very object of his own research and theory is likely a fiction, and yet he cannot resist the urge to continue with his endeavour. Canfield might be onto a good explanation why.

While Canfield's speculative natural history may help us achieve a Wittgensteinian perspective on the nature of the self, further results from primatology and evolutionary psychology can amplify his ideas to help us think clearly about the nature of psychological language and the mind. Evolutionary psychologists are deeply interested in how it is that we evolved to suppose that other people have minds at all, or to do *theory of mind* (ToM), a phrase that psychologists use differently from philosophers. For psychologists, ToM refers to the human capacity to effectively postulate other people's mental states in order to predict and influence their behaviour. For philosophers, the phrase refers to the unprovable claim that other people have private mental states at all. We will return to this distinction and its relevance shortly. Many or most evolutionary psychologists who write about ToM hold that humans have either an entirely unique capacity to explain and predict the behaviour of others or else one that is uniquely well-developed in the animal kingdom. Boyer and many other evolutionary psychologists assume that ToM evolved because it usually renders advantageous predictions of others' behaviour and insights that allow us to influence their behaviour. That is to say, this evolutionary-psychological hypothesis is that ToM is a reasonably good lay scientific theory constructed by the unconscious to predict and control things in the social environment in order to gain a reproductive advantage.

If ToM really is a theory of private mental phenomena, it is empirically unwarranted whether or not it is an unconscious compulsion. The philosopher's objection applies to postulates of unconscious systems as surely as it applies to the postulates of people. That is to say, behavioural evidence is always insufficient to prove the existence of private mental phenomena in others so long as we conceive of those phenomena as images, ideas, beliefs, etc. passing through conscious and unconscious spaces inside of the person, hidden from view. This truth is not lost on Justin Barrett, whose model of mind and whose critique of religion are virtually identical to Boyer's except for the particular mental systems that he proposes. Barrett writes that 'belief in minds is not empirically supported' and yet 'belief in minds is obstinately universal'.[37] This universal urge to illicitly posit other minds is blamed on the evolutionary pressure to quickly spot and understand other agents, other people who are the most challenging problems we face in our quest to propagate. 'Belief in minds arises from our species' biology working in the sort of environment in which we live', writes Barrett.[38]

In a truly remarkable moment of self-awareness, Barrett acknowledges that his own enterprise of describing the unconscious systems that are the focus of his book, along with the enterprise of most cognitive psychology itself, can be described as an investigation of that same thing – other minds – that is illicitly called into being by our animal unconscious. He writes, 'Ultimately, what psychologists study is human behavior, including the behavior of brains and nervous systems. Psychologists and cognitive scientists *interpret* behaviors in terms of mental states and the functions of minds. However, *minds* are not accessible to direct investigation and have not even been proven to exist.'[39] A few pages later, Barrett writes, 'Both belief in God and belief in other minds arise from the operations of nonconscious mental tools generating nonreflective belief.'[40] Ultimately, says Barrett, he cannot take the idea that other minds don't exist seriously because his animal nature compels him to believe otherwise. 'Belief in minds is *natural*.'[41] In other words, despite the evidence that belief in other minds is scientifically untenable, our animal natures present it so compellingly that Barrett cannot resist the temptation to make other minds the focus of his research.

Barrett's double-think is all the more remarkable given the premise of his book: our unconscious systems operate like little scientists, but they make mistakes and sometimes produce serviceable but false religious beliefs. Early in the book, he establishes his position that we can protect ourselves from the pseudoscience of the unconscious by invoking the real science of reflective conscious scrutiny:

> Reflective determination of plausibility is something that people rarely engage in unless they are formally trained to do so by scholars. Rather, they just 'feel' that the belief is sensible. Unless one is trained in logic or empirical reasoning, even when required to offer justification, whatever 'pops into mind' first may seem a good enough justification. This 'popping into mind' typically amounts to nothing more than a fragmentary reiteration of the nonreflective plausibility determination. . . . Indeed, much of the training in the social and natural sciences is teaching skepticism and how to find alternative explanations. It does not come easily.[42]

Why back down in the face of the apparent *reductio ad absurdum* of his research? Why give in to the mere feeling that other minds are real? The answer isn't clear.

The fact that people around the world do tend to posit other minds appears to be undeniable, but it is relevant to note that the particulars of their 'theories' vary substantially. Lisa Feldman Barrett notes:

> Not all cultures understand emotions as internal mental states. Himba and Hadza emotion concepts, for example, appear to be focused on actions. This is also true of certain Japanese emotion concepts. The Ifaluk of Micronesia consider emotions as transactions between people.[43]

On the same note, in his remarkable history of scientific psychology, Kurt Danziger explains that he gave up thinking of the mind as a thing whose nature could be discerned and described by science when he learned that in Indonesia, China, Uganda and other non-Western cultures, both laymen and psychologists posit entirely different mental functions and they work on entirely different problems. Were the findings of psychological science objective, he argues, surely there would not be such radical divergence around the world.[44]

Ironically, the evolutionary psychology of ToM may offer us a way out of the mobius illogic that brought us to this point, but we will have to separate its empirical findings from wayward theories of interior processes and belief structures such as those that Boyer and Barrett propose. What surely evolved in us are certain capacities and tendencies of *behavior*, broadly understood, and it is those capacities that underlie religious practices, the project of psychology, natural science and countless more human endeavours. The hominid capacities to do ToM and to use language appear to have evolved primarily due to the intense pressure of social life, not those posed by the wider environment.[45] Modern-day baboons, whose anatomy and natural environment are considerably more like those of our distant hominid ancestors than our own, remind us of the primary functions of ToM and speech. To survive and propagate, an individual baboon must remain acutely aware of the alliances and rivalries within her own troop and make 'nuanced decisions about when to join, whom to join, whom to threaten, and whom to ignore' at any given moment.[46] Through their copious research, the primatologists Dorothy Cheney and Richard Seyfarth furthermore demonstrate that baboon vocalizations are thoroughly integrated into their social contexts.[47] A successful baboon must master the arts of interpretation and expression that give any given coo or cry the meaning that it has in a particular case. She must track who said what to whom, as well as how, where and when it was said if she is to anticipate its effects on the troop and its significance for herself.

What Cheney and Seyfarth neither see nor say is that their research suggests that human evolution favoured the propensity and the capacity to *speak* about ourselves and others *as though* each person were ruled by a Cartesian ego. To navigate the social world, it is handy to frame and execute our political plans in a language that attributes systematic causes to human behaviour and that provides us tools with which to influence them. The brain and the rest of the human anatomy most certainly did evolve in ways that make it possible for individuals to acquire these language-games, these forms of life. However, there is no evolutionary imperative that our language, whose essential function is social, also describes a realm of hidden phenomena. If our *H. sapiens* ancestors spoke of *thoughts*, *ideas*, *dreams* and *hopes*, they were speaking about social facts, and it is a category mistake to suppose that they were – or that we are – really speaking about some peculiar processes hidden away beyond the reach of perception. Between the ears of monkeys and men there is nothing but matter: nerves, blood vessels, glial cells, etc. While it is useful to speak about others as conjuring memories, pondering options and feeling pains, the value and the meaning of all of these expressions can be cashed out in terms of their social significance. They are only so many linguistic behaviours integrated into our animal nature and our animal capacities that allow us to navigate our social environments. It is to this point that Wittgenstein speaks when he says:

> Believing that someone else is in pain, doubting whether he is, are so many kinds of natural behaviour toward other human beings; and our language is but an auxiliary to and an extension of this behaviour. I mean; our language is an extension of the more primitive behavior. (For our *language-game* is a piece of behaviour.)[48]

To interpret our talk about other minds as though it referred to actual hidden processes is to categorically misunderstand the logic of our language. Wittgenstein again:

> I want to regard man here as an animal; as a primitive being to which one grants instinct but not ratiocination. As a creature in a primitive state. Any logic good enough for a primitive means of communication needs no apology from us. Language did not emerge from some kind of ratiocination.[49]

Language, including our ways of speaking about other people's thoughts, did not evolve to articulate a theory or to describe theoretical posits. Indeed, the idea that such basic forms of language evolved around a complex theory of unprovable postulates is patently absurd. Language evolved to serve the human animal in its social and physical context, and both philosophy and psychology can only be written in that same language. When psychologists like Barrett and Boyer speculate about the interior processes of *memory, belief, thought*, etc., they are guilty of taking 'language on holiday',[50] that is, of using the terms that evolved to explain social facts as though they referred to some entirely different facts, viz., processes hidden from both perception and introspection, which is to say, wholly theoretical posits.

None of this is meant to say that understanding ourselves or others is easy or simple. The thrust of Wittgenstein's thousands of remarks on psychological concepts points in the opposite direction, towards the conclusion that our difficulties reading each other and understanding ourselves are a fundamental part of the human condition. Late in life, Wittgenstein wrote to Piero Sraffa:

> The older I grow the more I realise how terribly difficult it is for people to understand each other, and I think what misleads one is the fact that they all look so much like each other. If some people looked like elephants and others like cats, or fish, one wouldn't *expect* them to understand each other and things would look much more like they really are.[51]

And in his notebooks from the early 1930s, Wittgenstein makes the following remark:

> My self-knowledge is this: When a certain number of veils are left on me, I still see clearly, namely the veils. But if they are removed so that my gaze can get closer to my ego, my image begins to blur before my eyes.[52]

The challenges of understanding others and of knowing ourselves are inescapable parts of human life, and these difficulties are reflected in the complicated grammar of our socio-psychological language. Wittgenstein again:

> But is it not peculiar that there is such a thing as this reaction, this confession of intention? Is it not an extremely remarkable instrument of language? What is really remarkable about it? Well – it is difficult to imagine how a human being learns this use of words. It is so very subtle.[53]

Our ways of speaking about the mental lives of others and of ourselves did not evolve to describe the systematic workings of hidden mechanisms, they evolved as elements of the byzantine social patterns that are woven onto and into our animal capacities. Our talk about *beliefs, intentions, emotions*, etc. is as much a part of our animal origins and our animal lives as our capacities to sweat, to whistle a tune or to keep our eyes fixed on prey while we jog, and it is a category mistake to interpret them as names of interior phenomena. One might protest that the psychologist's aim is to describe the mechanisms that underly those animal reactions and capacities to which Wittgenstein himself alludes. Indeed, there are methods of doing that, but they are not the methods employed by Boyer. Though claiming to investigate hidden brain processes, virtually all of Boyer's hypotheses describe the sorts of psychological facts that actually exist only at a social level. They are pseudo-explanations of religious life born out of the idea that hidden internal systems can be literally described in a vocabulary that evolved for very different purposes as if the brain had beliefs and intentions rather than the person we recognize in our social environment.[54]

5.4 Meaning, ritual and thick description

Keeping in mind these lessons about the origin and nature of language, let us return to the topic of rituals. Boyer's chapter on rituals begins with brief accounts of four rituals, followed by the following remark:

> These examples were taken from Java, India, Nepal and Central Africa, but I could have chosen virtually any other continent or religion to illustrate the point.[55]

The fact that Boyer is not interested in the details of each ritual and its context is both jarring and instructive. It is no small matter to assume from the outset that all rituals from all around the world share one essential form. Boyer's explicit reasoning is that ritual behaviours are motivated by unconscious thoughts and the vague intuitions that bubble up from them to consciousness, and so they have no logic, no sense, that their practitioners know anyway. In fact, Boyer claims flatly that rituals simply don't have meaning:

> We often hear that ceremonies are *meaningful* to the people who perform them. Through ritual, people perhaps grasp important messages about themselves, their relationships to each other and their connection with gods and spirits. . . . But do rituals really convey much meaning? To be blunt, what does it *mean* to bite a ram's tongue or to get some mean-looking character to pretend he is killing your children? What is the information transmitted? Not much, apparently.[56]

The real motives for ritual are generated by unconscious processes to which a person has no direct access. The ritual participant is at best aware of only the 'vague intuitions' that bubble up from the unconscious into consciousness. Only the scientist can know the true origins and meaning of those intuitions, according to Boyer[57] since the subject has no direct perception of them in himself or in others.

Readers already familiar with Wittgenstein will recognize that Boyer's picture of meaning is the same one that Wittgenstein works so hard to liberate us from in the *Philosophical Investigations*. Boyer assumes that a *meaning* is something apart from the ceremony itself. He assumes that if a ceremony had meaning at all, that meaning would be something that accompanies the ceremony, it would be a piece of information conveyed to the mind of a ritual participant where it might then be restated in the form of a verbal assertion. It is because rituals have no meaning that people cannot explain their sense to others. According to Boyer, the fact that rituals don't travel with paired meanings explains 'why people's exegesis of their own rituals is often vague, circular, question-begging, mystery ridden and highly idiosyncratic'.[58]

Wittgenstein works assiduously to liberate us from this 'primitive idea' of meaning[59] by urging us again and again by way of countless remarks, aphorisms, thought experiments and the patient pursuit of misguided ways of thinking to their self-defeating ends. By whatever means he can muster, Wittgenstein gradually presses his readers towards a conception of language as activity much like the conception that emerges from Cheney and Seyfarth's primate research. Namely, language is a kind of human behaviour that develops as a component of a much larger, more encompassing family of acquired gestures built on and out of our instinctive animal ways of being. We naturally respond to the other's gestures of pain, and we gradually acquire the habit of asking 'where does it hurt?' We naturally wince with our own pain, and gradually learn to say 'Ow!' when we suffer small injuries. The meaning of an expression is its *use*, as Wittgenstein emphasizes over and over again, not a mental accompaniment. 'Hence it would be stupid to call meaning a "mental activity", because that would encourage a false picture of the function of the word.'[60]

Trapped by a false picture of meaning, Boyer embarks down the wrong path to uncover the sense of our rituals. If rituals convey no information, why perform them at all? Boyer's first, preliminary answer is that they are compulsive behaviours closely akin to OCD, an idea borrowed from Alan Fiske. The superficial similarities between ritual behaviours, which often prescribe specific sequences of action without any clear explanation why, 'suggest that some elements of rituals trigger activation of those particular mental systems that work in overdrive, as it were, in obsessive disorders'.[61] However, this explanation would seem to be more an irresistible slight than a serious proposal since Boyer goes on to provide another account that is wholly independent and largely incompatible with this one. Boyer's considered explanation of ritual is that it arises out of our unconscious intuitions to provide us ways to modify relations between people. 'Without clearly knowing why, the participants feel that the special "game" they are playing produces important changes in their relations. Which it does, although people could not readily explain how it works.'[62] Births, brisses, weddings and funerals are all rites that alter our social realities. Much as a baboon's social context can be altered by an interloper, a fight, a death, a well-timed grunt or well-received grooming, so too our context can be modified by successful ceremony. According to Boyer, our subconscious systems are performing 'precise calculations' about the social effects our actions will have like 'sophisticated game theorists', but we consciously experience only those vague intuitions mentioned earlier.[63]

Boyer's theory makes no real sense, however, since it borrows the terms of our evolved, socially governed language and takes them on holiday. Language evolved as part of our behavioural repertoire, and our words have no meaning or valid application that transcends their social reality. The concept of 'calculating' is as much a product of our evolved social behaviour as the words 'mama' and 'milk'. *Infants* want milk, not unconscious thinking machines inside them. So too, *rivals* calculate their odds of overthrowing the king, not unconscious processes. I don't mean to imply that our evolved intuitions and compulsions regarding social relations are irrelevant to the meaning of rituals. It may be that we are driven to alter our social relations by means we do not fully understand, much as Boyer says. My point is that Boyer's account of 'precise calculations' going on in a 'mental basement' is both categorically confused and, at least as importantly, profoundly misleading. If evolution produced in us compulsions to perceive social dynamics that we don't consciously grasp, then rituals and magic are valid solutions to solving problems posed by those compulsions. We might compare David Buss's evolutionary psychology of human sexual preferences and mating patterns. Buss often refers to our unconscious 'desires' and 'calculations' and the pressures they exert on our behaviour, but he never loses sight of the fact that these formulas exist in us *only* in the form of instincts, tendencies, urges, etc. Expressed otherwise, evolution calculated our best reproductive strategy, but what it handed off to our minds is nothing more than the tendencies that embody its results. Witness how Buss writes about the unconscious drives that influence married partners to stray:

> In evolutionary terms, it makes perfect sense that infertility and infidelity are the most prevalent causes of divorce worldwide. Both represent the strongest and most direct failures to deliver the reproductive resources that provide the evolutionary *raison d'etre* for long term mating. People do not consciously calculate that their fitness suffers from these events. Rather, infidelity and infertility are adaptive problems that exerted selection pressure on human ancestors for a psychology attuned to reproductive failures. Just as having sex tends to lead to producing babies although the people involved may have no awareness of the reproductive logic involved, so anger leads to leaving an unfaithful or infertile mate without requiring conscious articulation of the underlying adaptive logic.[64]

What evolution produced in us are the drives that we experience, the fears, the wants, the tendencies towards sex and violence and much else. It did not supply us with cognitive machines that unconsciously recreate the calculations of millions of years of evolution in the blink of an eye. To speak of such unconscious calculations may be a useful metaphor, but to treat it literally is terrifically confusing. The result, in Boyer's case, is a pseudoscience that undermines our attempt to understand ourselves and others better. The fears and desires that we experience are the brute facts upon which human forms of life are built; they are the foundation of meaning. One culture might build a legal code stipulating legitimate grounds for divorce to solve certain problems of human life while another culture might develop a system of rituals or magic to serve much the same purpose.

This is why Boyer's lack of interest in the particulars of the rituals he mentions is so galling; he cuts off our first and best hope of understanding what meaning those rituals have for participants. It may be hard for a westerner to understand why a Kham Magar shaman would bite a ram's tongue, but surely it is only decent to make an effort. My point is not that *all* rituals have meaning. I argue elsewhere that it is possible for some rituals to survive as vacuous customs,[65] much as Boyer suggests, but each case requires individual consideration. In a sensitive, ponderous essay about the challenge of understanding other cultures, Clifford Geertz reflects on the report of a nineteenth-century Dane, L. V. Helms, describing a Balinese ritual in which the dead king's three widows jumped to their deaths. Helms describes in some detail the eleven-story tower from which the women leapt, the great crowd of onlookers and the variety of offerings they carried, the gold and black sticks placed under the wooden image of a lion, a priest who shot a flower from a bow at a stuffed and colourfully painted snake. Helms is sensitive to the artistry and the complexity of the affair, even admiring of its ornament and pomp, yet he also reveals his own biases when he refers to the widows as 'deluded women' and to the whole event as a 'terrible spectacle'.[66] Geertz interprets Helms's complicated reaction as indicative of the profound difficulty we face when trying to make sense of something so alien and so frightful: on the one hand, we must strive to find our common humanity with the other and to see the world from her perspective. And at the same time, we can never fully transcend the frame of our own perspective, the concepts and values through which we see and describe the world. It is a kind of paradox, yet nonetheless it is possible to understand other people and their cultures, Geertz insists.[67] The process may be imperfect and plodding; we may need to double-back and think more than twice about what we bring to our account, but it can be done. Interestingly, Geertz includes this warning against the sort of psychological explanations that Boyer attempts:

> [The notion] of the material determination of consciousness . . . [may have some use] . . . but so far as understanding the construction of how other peoples' imaginations connect to those of our own, [it will] head us off in precisely the wrong direction – toward isolation of the meaning-form aspects of the matter from the practical contexts that give them life.[68]

The challenge is to imagine our way into the other's way of seeing the world so that we too grasp the 'meaning-form aspects' that he sees, so that the gestures and words he finds significant make sense to us.

To express Geertz's point in Wittgenstein's idiom, we must strive to see the other's context under the same aspect that he sees it. Weddings, divorce proceedings and infidelity hold meanings for us that any normal person (i.e. anyone who does not suffer serious emotional damage or brain damage) who participates in our forms of life can grasp even if he cannot verbalize them well. What does it mean to say that a mother baboon *understands* the import of her infant's plea for milk? It simply means that she responds to its gestures, both bodily and vocal, in some way that shows that she understands what is being asked of her. A mother baboon who consistently showed no response to her infant – not even a response that subtly signalled deliberate neglect – would be said not to understand her baby, and we might suppose that she suffered brain damage or profound emotional damage. There are criteria by which we judge whether a baboon understands the gesture of her baby just as 'there are certain criteria in a man's behaviour for the fact that he does not understand a word: that it means nothing to him, that he can do nothing with it'.[69] Baboons and people have animal natures that allow them to acquire complex forms of life into which are woven grunts, glances, social history, immediate context and more. A gesture or a word has meaning just in so far as it has a role to play 'in the stream of life',[70] to grasp a meaning is to know how to use that word or gesture and how to participate. Stanley Cavell expresses this crucial point regarding our animal capacity to take up meanings this way:

> To this human capacity for seeing or for treating something as something, Wittgenstein attributes our capacity for understanding, for what we might call the innerness of meaning we attach to words and gestures.[71]

In his book, *A Radical Pluralist Philosophy of Religion*, Mikel Burley both describes and demonstrates a method of understanding religions and religious practices by striving to imaginatively occupy the perspective of participants. More accurately, Burley proposes not a particular method but suggests that we use 'a plurality of methods, unconstrained by conventional disciplinary boundaries'.[72] We must be prepared to wander, to invent, to attend closely to details, and, ultimately, to look for insight wherever we can find it. Burley informally labels the sort of account that works 'thick description', a phrase he borrows from Geertz who himself borrowed it from Gilbert Ryle. The term is not a technical one and Burley is clear that we should not demand that it be defined precisely.[73] His point is that we must find descriptions that help us build up from those thin threads of our common humanity to which Geertz alludes towards an imaginative recreation of the other's 'way of being human'. Helms's account of the Balinese sacrifice and countless more examples from religions around the world remind us how hard it can be to acknowledge that 'human life can be like that',[74] as D. Z. Phillips says, that there are forms of life deeply unlike our own.

To make sense of cannibalism, a shocking and shockingly alien form of life, Burley notes that our first step must be to recognize that there are different cases and different varieties of cannibalism, each of which must be understood on its own terms. In the recent past, the Wari' of Brazil ritually ate both enemies killed in battle and their own deceased tribe members, but the two acts did not hold the same meaning for them. To make sense of the Wari' practices, says Burley, we must recognize that they did not conceptualize the distinction between animals and people as we modern Westerners do, and so the boundary between bodies and food was also different. 'In the case of warfare cannibalism', Burley writes, 'the bodies of slaughtered enemies would in fact be treated with *less* respect than those of animals killed for food. . . . [Thus] the very treating of them as food serves the symbolic purpose of denigrating and humiliating the victims'.[75] The Wari' ritual eating of tribe members carried another meaning: 'The consuming of the body was interfused with highly ritualized performances signifying the deep mutual dependencies between different members of the community'.[76] Thus did the Wari' use rituals to transform social relations.

Because Boyer imagines that inside the Kham Magar shaman there are cognitive systems doing their best to follow the logic of science (but sometimes misfiring and sometimes miscalculating, as primitive things will do), he is unable to grasp the sense of their forms of life. Wittgenstein criticizes Frazer for this mistake again and again, that is, the mistake of supposing that deep within all human beings we all hold the aims and values of Western science. 'What a narrow spiritual life on Frazer's part!' Wittgenstein erupts in exasperation. 'As a result: how impossible it was for him to conceive of a life different from that of the England of his time!'[77]. Our human nature and human experience pose a great variety of problems, each of which admits a great variety of solutions. We might put the point this way: once evolution settles on a strategy of providing us capacities and tendencies of pride, fear, love, lust, jealousy and so on, thereafter individuals and cultures must find ways of satisfying the problems of life. Wittgenstein speaks to this point in a remark targeting Frazer's scientism that applies equally to Boyer's:

> Burning in effigy. Kissing the picture of one's beloved. That is *obviously* not based on the belief that it will have some specific effect on the object which the picture represents. It aims at satisfaction and it achieves it. Or rather: it *aims* at nothing at all; we just behave this way and then we feel satisfied.[78]

Boyer is aware of this same mistake in Frazer, but he attempts to avoid it by moving the methods and aims of science out of consciousness and into hidden, unconscious systems. Boyer labels Frazer's mistake 'intellectualism', or the mistake of believing that people construct explanations for phenomena they don't understand and that 'primitives' (Frazer's word) hit upon false hypotheses. 'The mistake of [Frazer's] intellectualism was to assume that the human mind is driven by a *general* urge to explain. . . . Our minds are not general explanation machines. Rather minds consist of many different, specialized explanatory engines.'[79] The result of Boyer's simplistic move is predictably disappointing. His dismissive accounts of religion and religious life appear to be just as narrow-minded, just as colonialist in their judgements, and possibly even less able to help us truly understand religious ways of being human. To understand the religious lives of others, we must strive to see the world as they do, to see how their solutions to the problems of life work for them.

Notes

1 Barrett and Church 2013, 312 and see Atran 2002; Barrett 2004; Bering 2011; Bloom 2007; Boyer 2001. Guthrie 1993; McCauley 2011; Pyysiäinen 2009.
2 Wittgenstein 1993, 151.
3 Wittgenstein 1993, 137.
4 Wittgenstein 1993, 129.
5 Wittgenstein 1993, 139.
6 Boyer 2003.
7 Deutsch, 2010.
8 Boyer 2001, 215.
9 Boyer 2001, 177–91, 202.
10 Boyer 2001, 258.
11 Boyer 2001, 262.
12 Boyer 2001, 94–5.
13 Boyer 2001, 21.
14 Boyer 2001, 250.
15 Boyer 2001, 33.
16 Boyer 2001, 218–19.

17 Boyer 2001, 223.
18 Boyer 2001, 227.
19 Wittgenstein 1993, 127.
20 Wittgenstein 1993, 123.
21 Wittgenstein 1993, 127.
22 Frazer 1981, 13–19.
23 Wittgenstein 1993, 137.
24 Wittgenstein 1993, 131.
25 Wittgenstein 1993, 133.
26 Boyer 2001, 90.
27 Ibid., 201-2.
28 Boyer and Liénard 2006.
29 Wittgenstein 1993, 131.
30 Wittgenstein 1993, 123.
31 Canfield 2007, 37.
32 Canfield 2007, 40.
33 Canfield 2007, 45–8.
34 Canfield 2007, 48–57.
35 Canfield 2007, 45–57.
36 Canfield 2007, 80–96.
37 Barrett 2004, 95.
38 Barrett 2004, 96.
39 Barrett 2004, 96.
40 Barrett 2004, 98.
41 Barrett 2004, 104.
42 Barrett 2004, 28.
43 Barrett, 2017, 53.
44 Danziger 1997, 1–5.
45 See Tattersall, 2012.
46 Cheney and Seyfarth 2007, 64.
47 Cheney and Seyfarth 2007, 223.
48 Wittgenstein 1980a, §151.
49 Wittgenstein 1969, §475.
50 Wittgenstein 2009, §51.
51 McGuinness 2008, 450.
52 Wittgenstein 2016–, Ms-183:90-91. My translation.
53 Wittgenstein 1967, §39.
54 See Robert Vinten's chapter in this same collection for a further analysis of Boyer's category mistake.
55 Boyer 2001, 231.
56 Boyer 2001, 232.
57 Boyer 2001, 257.
58 Boyer 2001 233.
59 Wittgenstein 2009, §2.
60 Wittgenstein 1967, §20.
61 Boyer 2001, 240.
62 Boyer 2001, 246.
63 Boyer 2001, 250.
64 Buss 1994, 176.

65 Hoyt 2012.
66 Geertz 1983, 38–9.
67 Geertz 1983, 41–4.
68 Geertz 1983, 48.
69 PI §269.
70 Wittgenstein 1982b, §913.
71 Cavell 1978, 251.
72 Burley 2020, 2.
73 Burley 2020, 63.
74 Phillips 2007, 205.
75 Burley 2020, 131.
76 Burley 2020, 131.
77 Wittgenstein 1993, 125.
78 Wittgenstein 1993, 123.
79 Boyer 2001, 15.

6

Cognitive theories and Wittgenstein

Looking for convergence, not for divergence

Olympia Panagiotidou

6.1 Introduction

To write about the ways in which neurocognitive research and the philosophical school of Ludwig Wittgenstein can find a common ground for collaboration was a challenge for me,[1] given the fact that philosophical theories often support neurocognitive research findings, while modern cognitive theories sometimes update philosophical views and perspectives. Although my special area of research is history and particularly the history of the ancient Greek and Graeco-Roman cults, I have a special interest in neurocognitive approaches to religion, and to past religions in particular, and I have extensively combined neurocognitive research findings with philosophical theories in my research.[2] Therefore, I suggest that the examination of whether the philosophical work of one of the most significant philosophers of the twentieth century can complement and enhance neurocognitive research may have mutual benefits for both fields.

Writing this chapter, my intention is to highlight points of convergence, not of divergence, between neurocognitive approaches and Wittgenstein's philosophical considerations. In particular, I wish to show that interdisciplinary collaboration between philosophers, historians and cognitive theorists may be beneficial for the research in all these fields. In this framework I wish to trigger further discussions between neurocognitive scientists, cognitive theoreticians and Wittgensteinian philosophers.

6.2 Wittgenstein on the role of philosophy and science

Ludwig Wittgenstein was deeply interested in understanding human thought, and especially the ways in which humans make sense of the world – or otherwise the ways in which the world presents itself to humans. Although scholars have traced a significant alteration of his philosophical thinking from his early work, *Tractatus*

Logico-Philosophicus, which was published during his lifetime, to his later work, *Philosophical Investigations*, which was published after his death, throughout his life he attributed particular significance to language, the nature, role and limits of which he strived to grasp in his philosophical considerations.

In the *Tractatus*, Wittgenstein argues that the fundamental philosophical problems arise from misunderstanding the logic of language (*TLP* 4.003) and thereby the logic of thought and the world. In his view, thought consists of sensible propositions which represent facts, which are states of affairs that make up perceivable reality. What cannot be perceived and experienced lies beyond the realm of nature, that is supernatural or metaphysical and therefore nonsensical. It then follows that philosophical problems are nonsensical as well, and 'are in fact not problems at all' (*TLP* 4.003). Therefore, contrary to natural sciences, which observe and try to explain the perceivable world (*TLP* 4.111), the task of philosophers is to clarify human thought,[3] to spell out misunderstandings of the propositions – so language – that shape the world (*TLP* 4.05), and to point out the correct use of the words. So, engaging in philosophy comprises an activity per se, which does not intend to develop theories and doctrines. In this framework, he claims that philosophy is an activity that 'stands above or below, but not beside the natural sciences' (*TLP* 4.111). On the contrary, it may contribute to scientific research and philosophical theorizing, by pointing out when scientists – and traditional philosophers – suggest nonsensical propositions, misuse words and contradict the internal logic and limits of the world, thought and language.

Language gained more attention in the *Philosophical Investigations*, in which Wittgenstein transferred his focal point from the notion of logical forms that govern the world, thought and propositions to ordinary language, denying any kind of dogmatism. In his later philosophical thought, he considers that philosophy is not just an activity but a therapeutic activity, which intends to expose and heal fallacies in human thought, language, reasoning and world view. Away from any attempt to formulate firm principles, Wittgenstein attributed significance to contexts, which mediate the attribution of conventional meanings to words and thereby to propositions and language, which enable the development and unfolding of varied language-games in different circumstances. In his view, these language-games comprise a kind of human activity which is grounded on an underlying resemblance of particular cases (family resemblance) instead of definite rules and similarities which govern human communication. In this framework, words acquire meaning not by pointing out definite objects and facts, but through the ways humans use them to express concepts in specific contexts. Thereby, different contexts may impose different uses of the words, enabling the pairing between a word and its meaning in different language-games. That does not mean that words are invested with arbitrary meanings in different conditions. Both their contextual uses and the concepts which words express are 'in need of a justification which everybody understands' (*PI* §261). In this view, not an underlying logic but a network of commonly accepted rules determines which words and further which propositions and language-games make sense to the participants. This network of rules can be summoned up in the notion of grammar which makes linguistic communication a meaningful human activity (*PI* §§371, 373). However, Wittgenstein does not consider grammar as an external set of rules which should be learnt or is externally imposed to language-games. For him, grammar arises from people's forms of life and enables them to judge the intelligibility of both their own and others' propositions.

According to Wittgenstein, philosophers should be aware of the deceptive power of language and, 'since everything lies open to view there is nothing to explain' (*PI* §126), they should trace and expose the fallacies in the philosophical problems, making them disappear (*PI* §133). In this framework, their 'investigation is a grammatical one' and the intention is to clear up possible 'misunderstandings concerning the use of words' (*PI* §90). Therefore, the work and role of philosophers are quite different from those of scientists. While scientists observe the world, discover facts, conduct research, have research findings and develop theories about observed phenomena, philosophers should avoid any attempt to explain the world and to develop theoretical systems, being mainly interested in analysing concepts and pointing out misuses, fallacies and misunderstandings in scientific (as well as ordinary non-scientific) propositions. From this perspective, although philosophy and science have different objects of interest and purposes and follow different paths to accomplish their goals, their works can be supportive of each other, promoting a better understanding of human world.

6.3 – Wittgenstein and neuroscience(s): Phenomenal divergence and points of convergence

Although Wittgenstein drew quite clear boundaries between science and philosophy,[4] he has presented us with remarks about the distinct roles of these intellectual activities based on observations that seem to reflect fundamental attributes and processes in human cognition, which are currently studied and largely confirmed by neurocognitive research.

In the *Tractatus*, Wittgenstein attributes great significance to human thought as the gate which represents the world to humans. Human thought is articulated in sensible propositions which reflect and share the same logical form with the perceivable world. Human thoughts and therefore sensible propositions do not represent simple objects, but concatenations of objects and the relations between them which comprise possible states of affairs. States of affairs that obtain are facts that make up the world. In this view, simple objects are conceived as elementary perceptual units (e.g. colours, sounds, textures etc.) which are associated with each other to comprise certain states of affairs (i.e. combinations of elementary objects) which constitute the reality we perceive. Human thought and linguistic propositions may represent actual states of affairs being perceived in everyday interaction with the world, but may further represent possible states of affairs that do not obtain. However, in both cases what can be conceived and linguistically expressed in sensible terms coincide with what is possible to happen in reality, that is, with possible states of affairs.

Wittgenstein traces an internal connection between human thought, language and the world which attributes meaning and essence to the perceptible reality. This connection is being confirmed by neurocognitive research which throws light on the ways in which the external world acquires shape and becomes meaningful to humans. In particular, the surrounding world, both natural and artificial, consists of multiple electromagnetic waves (e.g. microwaves, sound waves, visible light, X-rays, ultraviolet rays, gamma rays) which carry energy and follow strict natural laws. This physical world, however, does not coincide with the world which is being perceived by humans, that is, the phenomenal world. The construction of the human organism determines and restricts which part of the wide electromagnetic spectrum can be perceived, attributing phenomenal shape to the world. Thereby, the human visual spectrum, for instance, contains wavelengths that range from 400 to 700 nanometres.[5] These waves comprise the visible light which is compatible with human visual system and is processed in the visual cortex of the brain, enabling humans to see and perceive shapes, colours and textures in their surroundings, and to discern objects, surfaces and backgrounds in three dimensions of length, width and height. These rudimentary percepts seem to correspond to what Wittgenstein considered to be simple objects. The perception and recognition of such simple objects derive from multi-modal elementary percepts which are combined with each other to generate different entities. Recurrent associations between elementary percepts and simple objects generate certain image-schemas which seem to correspond to Wittgenstein's states of affairs or facts. Such image-schemas (e.g. CONTAINER< BALANCE, CENTRE-PERIPHERY, FORCE-COUNTERFORCE) describe relations between simple objects and give rise to concepts that shape the external world[6] and underlie conscious thinking and linguistic communication.

In particular, multiple sensory stimuli, including visual representations, sounds, odours, flavours and textures are associated with each other or with specific mental and emotional states, such as feelings, emotions, desires, intentions and actions, and comprise image-schemas, which are kept in memory and enable humans to envision causal relations between them and potential future outcomes of currently perceived events or of their own current behaviours and actions. These image-schemas, which are implicitly stored in memory and 'comprise people's *implicit procedural knowledge*, develop into concepts during *perceptual meaning analysis* that "analyzes perceptual displays into meanings"'.[7] It is during the process of conceptualization, that the implicit image-schemas reach consciousness and become verbally expressed, so humans can explicitly think about them and verbally describe and manipulate them.[8]

It seems that Wittgenstein mostly thinks of this kind of *explicit conceptual knowledge*,[9] when he underlines the deep connection between human thought and linguistic propositions in *Tractatus*. It seems indeed that conscious thinking presupposes lexically articulated concepts and may develop even in the absence of immediate percepts to which the words refer. Thereby, humans are able to think of entities or events that happened in the past and to imagine things and states of affairs that may happen in the future. In any case, however, their thoughts can represent possible states of affairs, which consist of already stored image-schemas and concepts, and therefore, as Wittgenstein notes, the limits of – conscious – thought coincide with the limits of language and further of the – perceptible and conceivable – world.

As Wittgenstein moved from the inquiry concerning the logic that underlies and connects the world, the human thought and language in the *Tractatus*, to the examination of the role and significance of language for conceiving and attributing meaning to both personal and common experiences in the *Philosophical Investigations*, he explored how linguistic expressions and verbal communication become meaningful. He therefore concluded that humans share some common natural reactions to things which remain largely unchanged over time and in different cultural contexts, and which provide the ground for the development of language and for verbal communication. As natural human reactions are invested with meaning and become verbally expressed they may be modified affecting human behaviour. Therefore, as Wittgenstein aptly puts it, 'speaking of language is part of an activity, or of a form of life' (*PI* §23). A genuine and accessible product of this form of life is grammar, which Wittgenstein conceived as a set of rules which determines the meaning and use of the words, the structure of sentences and the conceptualization of human perceptions and actions, making language a meaningful activity (*PI* §371, 373).

These are crucial remarks made by Wittgenstein which preceded and echo the more recent inquiry of neurocognitive scientists concerning the origins of human language. In particular, neurocognitive research sought these origins both in the phylogenetic and ontogenetic development of human beings. If Wittgenstein with the notion of 'form of life' meant a common way of living, perceiving and conceptualizing the world shared by humans, neurocognitive research located this shared way in human cognition. And similarly, if Wittgenstein found in grammar the underlying norms which structure human thought and verbal expression of thinking and communicating these thoughts, neurocognitive research found the deepest roots of this grammar in the cognitive abilities which humans developed phylogenetically and individually acquire during ontogenetic development. In this view, grammar is the phenomenal and thereby observable structure and expression of the ways in which humans perceive, conceive and conceptualize the world in the ways they do because they have these specific neurocognitive abilities to attribute shape and meaning to phenomenal reality. And these shared neurocognitive abilities enable humans to develop and display common natural reactions and 'forms of life' which allow them to use language as a mode of communication between themselves.

I would dare to say that from both a Wittgensteinian and a neurocognitive perspective, meaning is not an inherent property of the words neither of the objects to which the words may refer. Words – as well as other potential vehicles of symbolic communication (e.g. facial expressions, gestures, objects, practices) – acquire meaning by being used in specific contexts, and this meaning may alter or be modified when these contexts change as well. What, however, remains and makes it possible for people to understand words, to interpret language and to attribute meaning to perceivable world is an underlying system of reference as it is expressed in grammar and lived in humans' shared forms of life, and which derives from common cognitive abilities for symbolic communication developed by humans throughout evolution.

In this light, Wittgenstein's insights into both the logic that connects human thought, language and the world in *Tractatus* and into the shared forms of life and grammar as the modes that underlie human ability for making sense of the world and attributing meaning to the words in the *Philosophical Investigations* seem topical and in tune with current neurocognitive findings.

6.4 Seeing versus seeing as: Neuronal plasticity and contextualization

Conceptualization comprises a crucial semiotic process during which perceptions and implicit image-schemas are transformed into explicit concepts. Wittgenstein paid particular attention to the final products of this process, namely to the concepts used by people in both thought and communication. Neurocognitive research, on the other side, is more interested in exploring the processes that enable humans to perceive, conceive and conceptualize the sensory stimuli received from their surroundings. The different – but complementary and not contradictory – orientations of these streams of inquiry can be summarized in the distinction between 'seeing' and 'seeing-as', which Wittgenstein suggested and explored in the *Philosophical Investigations.*[10]

For Wittgenstein, 'seeing' refers to visual perception, while 'seeing-as' is related to a visual stimulus which is being visually perceived and simultaneously conceived and interpreted as being something. The sensory process of seeing demands complex neuronal and cognitive mechanisms in place. The interpretational operation of 'seeing as', however, is an even more multivalent process, which involves the neurocognitive apparatus for sensing – seeing and so perceiving – the world along with the situatedness of this apparatus in specific contexts which affect the processes of conceptualization, interpretation and meaning-attribution to purely sensory stimuli.

In this view, language concentrates and reflects the whole range of processes which makes the 'seeing as' possible. From the moment of birth, humans live and grow into specific linguistic contexts in which they learn to understand, recognize, categorize and express their perceptions into concepts and further to make sense of both their private and shared experiences. Thereby, human experiences are being interwoven with language which develops into the main instrument for shaping their thoughts and understanding both themselves and the world.[11]

Recognizing the significance of language for making sense of both individual and shared perceptions and experiences, Wittgenstein rejected the notion of 'private language', considering that individuals may perceive or feel, recognize and categorize even the most inner sensations and feelings using words from the language into which they grew up. And further, although other people may not share the same sensations and feelings with another person, they are able to recognize, categorize and conceive the others' feelings and sensations by observing those people's actions and reactions, and by drawing the appropriate terms from the common pool of concepts and words.[12]

In this light, Wittgenstein claims that philosophers should not strive to explain human experiences and the physical world. They should give up any attempt to formulate empirical hypotheses and theories and to provide explanations, and focus on 'the workings of our language' (*PI* §109) in order to describe what is already known in meaningful terms.[13] They should mainly focus on describing how these experiences and the world feel to humans and furthermore how these common feelings and sensations provide the ground for communication and shared views of the world.[14] In this framework, the role of philosophers is not to explore how and why humans developed the ability for language and social communication, but to describe the language-games in which words are interrelated with their uses in different contexts and thereby become meaningful for people who participate in these games and interact with each other and the external world.

Wittgenstein attributed great significance to the contexts in which words are used and may be invested with different meanings. The role of contexts for the development of human cognition has been also recognized by modern cognitive theories – theories that mainly develop in the field of the study of religion.[15] In particular, although early neurocognitive research seemed to overestimate the human brain in relation to other components of complex human beings, soon researchers came to recognize that human cognition does not just coincide with the neuronal processes that take place in the brain, but develops through its embodiment in the human body and – via this embodiment– through its situatedness in the surrounding world.[16] This recognition of the intrinsic connection between the human brain, the body and the world – both natural and cultural – underlines the role of contexts for 're-engineering' human cognition.[17] In this view, human cognitive capacities are not fixed qualities that enable humans to perceive themselves and their surroundings and to interact with others and the world in a standard and predetermined fashion. On the contrary, embodied human cognition continually interacts with the contexts in which it is situated and which may touch off adaptations, modifications and evolution of human cognitive abilities.[18]

These insights into human cognition entail that similarly to words and thereby to language, cognitive abilities are not already installed in the human brain and being used as such by humans in their everyday interactions with their surroundings. During evolution, humans have developed multiple cognitive abilities – amongst which symbolic thought and language are included – which enable them to conceive and conceptualize their embodied experiences of both themselves and the world. These abilities are activated and developed in specific contexts which not only form the contents and external manifestation of cognitive processes but may also affect and modify these processes per se. In this view, Wittgenstein's crucial observation that grammar is the common ground of all languages which sets the rules for meaningful verbal communication reflects the neurocognitive insight that grammar is the genuine manifestation of the human cognitive ability for linguistic thought and communication, while the words, propositions and their meanings comprise the dynamic products of the situatedness of human beings in the specific contexts which form the contents of language.

The neuronal plasticity and the dynamic interrelation between sensory stimulation, perception and conceptualization do not entail that humans may have purely individual, private experiences. Wittgenstein supports his idea by arguing that people, who grow up and live in a particular context, share common forms of life which largely influence the ways in which they view the world and conceptualize their experiences, using linguistic tools and practices common to that context. Neurocognitive research further suggests that the embodied experiences that humans may have largely depend on and are determined by their neurocognitive apparatus, which further affects the forms of life and the modes of linguistic communication that may develop in different contexts. These insights imply that even the most private experiences that people may have cannot transcend the limits of their bodily and neurocognitive abilities and cannot be conceived and conceptualized out of the contexts of their lives and their linguistic communities.

6.5 History, philosophy and natural sciences: From opposition to collaboration

So far I've pointed out some points of intersection between Wittgenstein's philosophical thinking and neurocognitive research. Both streams of inquiry emphasize the role of contexts in the formation of human experiences, and this is where the field of history comes to the fore. Contexts alter over time, affecting people's forms of life and thereby language. Varied historical schools study the changes and alterations in contexts and the factors (e.g. significant events, turning points, significant personalities) that motivated these changes, and construct narratives about the course of alterations and their effects on people, communities and the surrounding world.[19]

Wittgenstein's approach to philosophy seems to echo Spengler's understanding of the historians' work.[20] Spengler[21] considered that history could not be understood by merely putting together the historical facts and events in a narrative and by explaining them in terms of strict laws (e.g. causal laws).[22] Historians should furthermore trace 'morphological relations', connections and analogies in seemingly different conditions and facts as well as between significant personalities in order to grasp the unities that underlie human history.[23] Wittgenstein transferred the notion of morphology to philosophy and suggested that philosophers should look for connections, commonalities and analogies between words, notions, concepts and their uses in language-games in order to better understand the human 'forms of life' and to dispel philosophical problems. This conception of philosophy does not provide new information, does not discover new things and nor does it generate new knowledge or theories. It mainly offers a new way to look at things which are already there and you are aware of them. In Wittgenstein's terms, 'it characterizes the way we represent things, how we look at matters' (*PI* §122[24]).

Therefore, from a Wittgensteinian perspective, philosophy does not comprise a cognitive discipline, does not make empirical discoveries and so does not resemble to natural sciences. This however does not mean that philosophical studies are less important than scientific research. On the contrary, while sciences study the natural world, discover facts and provide new knowledge about physical reality, philosophy clarifies what is not clearly defined, spells out conceptual ambiguities and provides a new understanding of what is considered to be already known (*PI* §109; Beale 2017).

Wittgenstein's attempts to identify the distinct roles of scientists and philosophers, and to separate philosophy from natural sciences, reflect his concerns about the overestimation of sciences at his time, which could lead to devaluation of other inquiries and rejection of other disciplines – such as philosophy and history – as being obsolete.[25] Noticing the dangers that may arise for philosophy from adopting the scientific methods of research, Wittgenstein managed to reveal the value of philosophical inquiry for natural sciences: philosophers may help scientists by analysing their scientific propositions, by scrutinizing theoretical articulations, by finding obscurities, misconceptions and confusions in the conceptual structures and linguistic expressions in which scientific knowledge and theories are articulated. This could be summarized as follows: natural sciences discover new facts, connections and laws in the world, philosophy helps better articulate and conceive scientific discoveries. If history is added in this schema, then scientific knowledge, conceptually refined by philosophy, can be monitored and better understood on a long-term scale, being tested in different contexts that extend all over the history of human culture.

6.6 – Applying cognitive theories to historical material

Studying history, I attended many classes about the different historical periods and the major events and political, economic and social developments that marked each period and mediated the conventional temporal division of the otherwise united human past. The significant personalities who participated in those events and affected the course of history are mainly mentioned in historical accounts and are pointed out for their contribution to historical developments (e.g. Pericles, Alexander the Great, Napoleon). However, ordinary people – although they do not usually draw attention – are those who live the events taking place in each era, are affected by the political decisions and the economic developments, participate in cultural practices and formulate history. This realization urges historians to seek insights into ordinary people's lives and to explore their thoughts, emotions, actions, behaviours and overall experiences of the unfolding events, which mediate the changes and transitions in the historical course.[26] The historical interest in everyday people's experiences has led to psychological interpretations of past people's mindsets and has generated historical schools (e.g. the School of Mentalities which developed within the Annales School at the beginning of the twentieth century, Psychohistory, Psychobiography), studies and approaches which pay particular attention to past people's belief systems, perceptions, collective attitudes, psychological expressions, modes of thinking, systems of symbols and world views.[27] There are certainly shortcomings in these approaches and certain objections have been raised to the possibility of studying the psychological dispositions, the mental and emotional states and the modes of thinking of past people. There are long philosophical debates about these very terms and how we can conceive and appropriately use them. There is also the historians' scepticism about the possibility of gaining insights into past people's thoughts and emotions and the impossibility of doing fieldwork in the same ways that anthropologists do. The interesting point, however, is that even Classicists and historians who reject the transfer and use of terms, notions and theories from other fields (e.g. from psychology) which deal with human minds and emotions to historical studies cannot avoid using psychological terms and references to mental and emotional states when they come up to talk about everyday people in their historiographical works.[28] And the main reason for that is that historians, as all humans do during thinking, reasoning and argumentation, widely use a set of principles, which are shared and are being accepted as self-evident in everyday communication, in order to articulate historical narratives and 'fill in the blanks in historical accounts'.[29] These principles comprise our so-called common sense, which has deep roots in the ways we think, reason and conceive ourselves and the world. And it was at this point that I found that neurocognitive research findings and theories about human cognition could be embedded in and support historical research, by providing new evidence, testable and supported by scientific methods, which would shed new light on the historical interpretations based on common sense.

The need to find theoretical and methodological tools to historically study the private experiences of past people without just reference to common sense became more urgent for me when I came to study two religious systems of Graeco-Roman antiquity, the flourishing and diffusion of which were mainly dependent on the attraction of new supplicants and adherents. The first such religious system was the cult of Mithras, a mystery cult which spread throughout the Roman Empire (except Greece). The second one was the cult of Asclepius which developed into the most significant healing cult of Graeco-Roman antiquity.

Adherence to Mithras presupposed initiation into a mystery cult, which derived from the initates' personal choice and not from any external impulsion, and participation in closed, exclusivistic communities of adherents, in which the initiates inhabited new roles and identities prescribed by the mystery context.[30] Worship of Asclepius was even more individualistic, in the sense that Asclepius was a divine physician who healed the sick. A person, who was afflicted by an illness or a disease, could turn to him, visit an asclepieion and ask for a cure during a ritual called incubation.[31] Both cases, initiation into the Mithras cult and participation in the incubation, were matters of personal initiative and engendered individual, personal and private religious experiences to the adherents. Therefore, for studying such religious traditions, it is not enough to delineate their historical evolution over time, to talk about significant personalities which might have influenced this evolution, to outline the cultural and symbolic systems of these traditions or to describe their practices without reference to the people who exerted and participated in them. And the major question which arose was how I could refer to individuals and the ways they experienced their participation in the religious rituals of the Mithras cult as well as of the Asclepius cult. At this point, neurocognitive theories seemed to me a good chance to find the theoretical and methodological tools to approach and better understand the people who lived in Graeco-Roman antiquity and took the decision either to be initiated into the Mithras cult or to visit an asclepieion.

The first step was to find out how I could conceive those people as distinct entities and then how I could trace their experiences, and what may be defined as experience after all. The new conception of human beings, suggested by neurocognitive sciences, provided a well-articulated theoretical framework for my research. One of the major objections to cognitive sciences raised by philosophers is condensed in the argument that cognitive scientists attribute psychological predicates to the human brain, saying, for instance, that the brain is thinking, dreaming, inferring and so on.[32] Such expressions partly derived, and could be excused, from the enthusiasm of the first neurocognitive scientists who managed to gain access via cutting-edge technological tools[33] and to look into the brain, an organ which had largely remained hidden until then. The technological advances in neuroimaging techniques and the new opportunities that opened up seem to have indeed prompted neurocognitive scientists and theorists to overestimate human brain over other components which compose complex human beings. However, one of the most significant developments in the field of neurocognitive sciences in the last twenty-five years was 'the realization of the intrinsic connection between human cognition and culture', which generated new theories for better conceiving multifaceted human entities.[34] Amongst these theories, the biocultural theory of religion suggested by Armin Geertz (2010) provided me the theoretical scaffolding to approach and try to conceive the experiences of people of Graeco-Roman antiquity.

Geertz's bio-cultural theory (2010) points out the inextricable connection between various elements which compose human entities. I need to clarify here that in this context human entities are conceived as multifaceted entities who are self-aware and aware of the external world, which they can distinguish from themselves and may interact with their surroundings. From this perspective, Geertz's theory (2010) suggests that humans must be first conceived as '*embrained* and *embodied* organisms'.[35] That means that human brain is embodied in a body and cannot operate outside of or without a body. The body and thereby the embodied brain are situated in a specific world. These three elements, the brain, the body and the world 'constitute a complicated interactive system, comprising of numerous subsystems', which mediate the function and deployment of human cognition.[36] In this light, the brain is not considered to be the seat of cognition but a crucial centre of coordination of cognitive functions. This view is supported by the organization of our nervous system, which includes the central nervous system, mainly consisting of the brain and the spinal cord, and the peripheral nervous system, which extends to the brain and our whole body and connects cognition with the external world.

Geertz also talks about the kind of connection which develops between the external world and human cognition. In particular, he argues that human cognition, in addition to being embrained, embodied and situated in the external world, is further 'extended, distributed and often imprinted' on the external natural environment as well as on the social and cultural surroundings.[37] In this light, natural objects and artefacts as well as conceptual symbolic constructions may develop and serve as 'cognitive anchors' to which individual cognitions may be extended and imprinted.[38] Every material object or conceptual construction (e.g. language, symbolic systems, cultural artefacts, practices) may mediate the extension and distribution of individual cognitions which are not constrained to brains and may facilitate communication between members of a cultural group or community. In this framework, the psychologist, neuroanthropologist and cognitive neuroscientist Merlin Donald (2001) has defined any given culture not as an external 'set of shared habits, languages and customs' but as 'a gigantic cognitive web, defining and constraining the parameters of memory, knowledge, and thought in its members, both as individuals and as a group'.[39] Summing up in a few words Geertz's bio-cultural theory (2010), human cognition is embrained, embodied and situated in the external world, to which it is extended and distributed. During the incessant interaction between human beings and the external world, natural objects and cultural constructions may 're-engineer' cognition in a process of continuous evolution of the cognitive capacities of the human species.

Using Geertz's theory as theoretical framework for my research, I approached the initiates in the Mithras cult and the supplicants of Asclepius as embrained and embodied entities who were situated in the Graeco-Roman world, living in which they interacted with each other and their cultural surroundings, and all these elements influenced their decisions and experiences during their participation in the relevant religious practices. In this framework I was interested in exploring the processes and functions which would have taken place in the brains and bodies of Mithraists and of the Asclepius supplicants and which would have been probably influenced and modified 'by every alteration or stimulus coming from their internal or external milieu'.[40] Relying on the theoretical premise that ancient people shared common cognitive mechanisms and processes with modern humans, I took into account findings from neurocognitive research and I attempted to explore how the specific social and cultural contexts of the Graeco-Roman world could have interacted with these processes re-engineering those people's cognitive mechanisms and abilities. In particular, I explored how the natural environment, the material artefacts, the architectural constructions, the pieces of art, the symbols and the language used in the cult contexts could have not only influenced the contents of the initiates' and supplicants' cognition, but could have also operated as 'cognitive components' which would extend and distribute their cognition and 'intelligence beyond the borders of their biological brains'.[41] In addition, I explored the ways in which 'collective emotional and conceptual networks', which developed in the wider cultural contexts of the Graeco-Roman world as well as in the specific religious context of the Mithras and the Asclepius cult, operated as powerful 'cognitive governance systems' which decisively affected the ways in which the initiates and supplicants perceived themselves, their position in the world and the others and shared coherent systems of worldviews.[42]

I selected the bio-cultural theory of religion developed by Geertz (2010) for my research because I found that it offered a complete conceptual framework to somehow approach the individuals who lived in a past era, Graeco-Roman antiquity, and to gain some insight into the private experiences of those individuals during their participation in specific religious practices. This view of human cognition is far away from any form of Cartesian dualism between mind and body or between brain and body implied by the first neurocognitive scientists. On the contrary, it provides a holistic view of human cognition and highlights the essential role of every part, the brain, the body and the world, for the development and employment of human cognitive abilities.

Try to think of yourself without one of these components. Think of yourself without a body. What remains is a brain and a world, but no portal, no connection between them. The body situates us in a specific environment. In addition, we constantly receive information from our surroundings through the sensory organs of our bodies, and the peripheral nervous system continually sends information to our brains which process this information and mediate our embodied interaction with the world.

Now think of yourself without the brain. You are a body situated in the world from which you receive multiple sensory information. This information, however, is not getting processed anywhere.

And now think of yourself as being out of space and time, out of any kind of perceptible surroundings. Like you got born in a dark place without any kind of sensory input from external surroundings. You have a body and a brain but no inputs to process.

In this light, the brain as a biological organ is necessary for human cognition to deploy, but this would not happen without a body. But again having a brain embodied in a body, but without being situated in specific external surroundings cognition would not deploy. Therefore, I will underline, once again, that the brain, the body and the external world comprise an interactive system which enables human cognition to develop and evolve both ontogenetically and phylogenetically.

6.7 Concluding remarks: Towards interdisciplinarity

From all natural sciences, cognitive neurosciences are those which study human cognition. Although – as already mentioned – early studies mainly focused on the neuronal networks in the brain, the realization of connection of those networks with the whole body and the external world, opened new paths for research which explore the ways in which human cognition develops, evolves and deploys in interaction with its surroundings.

Specific cognitive abilities, which developed responding to challenges and demands of both the natural and communal world, including the ability for symbolic thought, generated language and verbal communication. Language is closely connected with conscious thinking and conceptualization of time, abilities which further attribute narrative structure to embodied human experiences.

In particular, the ways in which humans perceive and conceptualize time (i.e. past-present-future, before-now-after) along with their mnemonic abilities (e.g. recall of personal memories and of stored information and knowledge) enable them to construct narratives in which they narrate their short-term life-stories as well as the long-term history of the human species and of the world. Conception of time, memory systems and narrative construction of human experiences are amongst the specific abilities of human beings that gave rise to the notion of history and to the development of historical studies.

Symbolic thought, perception of time and linguistic abilities are closely associated with other higher-order cognitive functions, for example conceptualization, causal thinking, inferential thinking, evaluative thinking, planning, decision-making and so on, which made humans capable of developing sciences as a systematic way to acquire and accumulate knowledge about the world, to explain the observed phenomena, to test the suggested explanations and to make predictions about the future of these phenomena.

Humans are able not only to observe, speculate and explain perceptible reality and empirical facts. They further become aware of their own awareness, think about their own thinking, speculate on what they know and also control and monitor the process of their cognitive endeavours. These are higher-order thinking skills which are described by the term 'metacognition'.[43] In this framework, philosophical thinking, in the ways in which Wittgenstein conceives it, comprises a high-order metacognitive ability. Philosophers speculate on the already accumulated knowledge, they analyse the linguistic forms and test the grammatical correctness of the propositions in which this knowledge is articulated, and clarify the conceptual fuzziness that may arise from misuses of the rules of grammar.

Based on those remarks, I suggest that historical studies, scientific research and philosophical thinking comprise complex mental activities which are deeply rooted in the cognitive and metacognitive abilities that humans develop throughout the evolution of their species. Although these activities have been organized into distinct academic disciplines and have developed varied theories, methods and practices for accomplishing their purposes, their roles seem to be complementary. Scientific research intends to explain the perceptible world, but the suggested explanations would lack any diachronic validation if they were constrained to the present without any reference to evolution and developments throughout history. Historical studies should cover any field of human culture. The history of scientific research and accomplishments could not be excluded. And since both scientific and historical studies are articulated in linguistic propositions, philosophical analyses of those propositions could ensure clarification of any fallacies and a better understanding of what is being said. And all these studies can be better and more deeply understood, if research findings on human cognition are taken into account, which can further throw light on the underlying mechanisms of conceptualization and conceptual fuzziness that may affect verbal communication.

In this light, interdisciplinary collaboration seems to be the new ground on which scholars from different disciplines may promote their research. In particular, by joining their methods and accomplishing their distinct roles, researchers from the disciplines of philosophy, history and cognitive sciences may deepen their knowledge and understanding of both human beings and the world. By neglecting each other, they might develop obsolete approaches and interpretations, lacking any universal value and sinking into conceptual fuzziness and confusion. Then, what remains is for scholars from these fields to realize the value of interdisciplinary collaboration, put aside their divergences and look for convergences.

Acknowledgements

I would like to thank Robert Vinten, Sofia Miguens and Nuno Venturinha for the invitation to participate in the third ERB Workshop: Wittgenstein, Religion, and Cognitive Science and to write a chapter for this volume.

Notes

1 I am grateful to Robert Vinten for sending me some material about Wittgenstein's thinking and philosophical considerations as well as about the main objections which are raised towards cognitive theories from a Wittgensteinian point of view.
2 For example Pachis and Panagiotidou (2017), Panagiotidou with Beck (2017), Panagiotidou (2022).

3 'Philosophy aims at the logical clarification of thoughts' (*TLP* 4.112).
4 'To say nothing except what can be said, i.e. the propositions of natural science, that is something that has nothing to do with philosophy' (*TLP* 6.53).
5 See, for example, Simos 2002, 27; Foley and Matlin 2015 [1991], 44–6.
6 Burgess 2006, 557; Evans 2010, 39–40.
7 Panagiotidou with Beck 2017, citing Mandler 2006, 67.
8 Panagiotidou with Beck 2017, citing Mandler 2006, 64.
9 Mandler 2006, 67.
10 Dunlop 1984; McGinn 2015.
11 Campbell and O'Sullivan 2015; Schroeder 2006, 201–19.
12 See Schroeder 2006, 201–19.
13 'And we may not advance any kind of theory. There must not be anything hypothetical in our considerations. We must do away with all explanation, and description alone must take its place. And this description gets its light, that is to say its purpose, from the philosophical problems. These are, of course, not empirical problems; they are solved, rather, by looking into the workings of our language, and that in such a way as to make us recognize those workings: in despite of an urge to misunderstand them' (*PI* §109).
14 See Dunlop 1984 and Schroeder 2006, 201–19.
15 See, for example, Varela, Thompson and Rosch 1991; Damasio 1994; Clark 1997; Geertz 2010.
16 Geertz 2010; Pachis and Panagiotidou 2017.
17 Wilson 2010, 180–1; Geertz 2010, 310.
18 cf. Geertz 2010.
19 Pachis and Panagiotidou 2017.
20 Spengler 1918; Beale 2017.
21 Spengler 1918, 49.
22 Beale 2017, 64.
23 Beale 2017, 64; cf. Miller 2009, xvi.
24 Cited by Beale 2017, 64; cf. Monk 1990, 302–3.
25 Beale 2017, 69.
26 Pachis and Panagiotidou 2017, 90.
27 Pachis and Panagiotidou 2017.
28 Dodds 1951, esp. 106, 114, 116, 120, 123, 134, 151–152; Burkert 1985, esp. 3–4, 58–59, 142, 190, 199, 220; Versnel 1981, 2011 e.g. 148, 170, 217, 259; Bremmer 1994; Parker 1998; see Pachis and Panagiotidou 2017, 91.
29 Pachis and Panagiotidou 2017, 93.
30 Panagiotidou with Beck 2017.
31 See Panagiotidou 2022.
32 In the *Philosophical Investigations* §281 Wittgenstein remarks: 'only of a living human being and what resembles (behaves like) a living human being can one say: it has sensations; it sees; is blind; hears; is deaf; is conscious or unconscious'.
33 For example, positron emission tomography (PET) and functional magnetic resonance imaging (fMRI). The PET scan measures blood flow in the brain. 'PET is a scanning technique allowing for the recording of changes in the brain after injury with or without cognitive activation (Tate, Shenton, & Bigler 2012) . . . Functional MRI is an imaging technique that detects which brain areas are involved in neural actions, cognitive processes, or emotional responses. It is also a widely used method to investigate diseased parts of the brain' (Billings, Churchill and Payne 2010).

34 Pachis and Panagiotidou 2017, 89.
35 Geertz 2010, 306–8; Panagiotidou 2014, 17.
36 Panagiotidou 2014, 17.
37 Panagiotidou 2014.
38 Panagiotidou 2014, 17.
39 Panagiotidou 2014, 17, citing Donald 2001, xiv.
40 Panagiotidou 2014, 17, 2022; cf. Geertz 2010, 306–8.
41 Panagiotidou 2014, 18, citing Geertz 2010, 310; cf. Norman 1993; Day 2004.
42 Panagiotidou 2014, 18; cf. Geertz 2010, 311–13; Donald 2001.
43 Flavell 1976.

7

Wittgenstein, naturalism and interpreting religious phenomena

Thomas D. Carroll

7.1 Introduction

'Is religion a natural phenomenon?' Thinking over this question with Wittgenstein in mind, I am tempted to say that the question is rather strange; its concision hides complexity. Much depends on how we understand what it is for something to be religious or natural. One might speak of an animal's behaviour as being natural, as being consistent with what has been previously observed or is otherwise readily explainable. One also might speak of a person's death as stemming from natural causes, as being the result of an internal disease and not an external factor. Furthermore, one might speak of certain sorts of feelings such as anger or compassion being natural under the right circumstances. To speak of something like religion as being natural, what may be meant is that religious beliefs (or perhaps more narrowly, beliefs in supernatural agents) hypothetically derive in some non-veridical way from the otherwise proper functioning of the neural systems of human beings, whether as an adaptive cognitive propensity or as a by-product of otherwise adaptive cognitive systems.[1] Some naturalist critics of religion, like Daniel Dennett, argue that such evolutionary explanations undermine the warrant of beliefs in supernatural agents.[2] Frequently at stake in philosophical debates over naturalism in philosophy of religion is whether religious beliefs (again, typically understood as beliefs in supernatural agents) ought to be appraised *epistemically*, that is, as genuine candidates for knowledge,[3] rather than *psychologically*, that is, as puzzling phenomena to be explained rather than taken seriously in their own right as putatively veridical states of mind.[4]

In what follows, I explore some ways in which Wittgenstein might be taken to support and alternately oppose naturalist approaches to interpreting religious phenomena. The chapter has three parts. The first part provides a short overview of some passages from Wittgenstein's writings – especially the 'Remarks on Frazer's *Golden Bough*' – relevant to the issue of the naturalness of religious phenomena. The second part explores some possibilities regarding what naturalism might mean in connection with Wittgenstein. Lastly, the third part considers the bearing of Wittgenstein's remarks on religion for the interpretation of religious phenomena. Ultimately, I argue that Wittgenstein's remarks on religion exemplify a way of thinking about the naturalness of religious phenomena, and that naturalistic depiction is part of the clarificatory work of philosophy. Wittgenstein reminds himself and his readers that religiosities are not themselves mysterious, per se; they point instead to what is possible within human life.

7.2 Wittgenstein's remarks on religion

Wittgenstein's remarks on religion are spread across a perplexing array of sources, and the remarks have produced a wide range of philosophical interpretations. Perhaps contributing to the interpretive puzzles are the different sensibilities Wittgenstein has across his corpus when it comes to considering religiosities.[5] Gorazd Andrejč has identified at least four 'conceptions' of religiosity in Wittgenstein's remarks on religion: what he calls the 'grammaticalist, instinctivist, existentialist and nonsensicalist conceptions';[6] I say 'at least four' because Andrejč observes that this list should remain open-ended.[7] The *grammaticalist* conception holds that statements of religious doctrine establish the rules of expression of beliefs for a religion. The *instinctivist* conception holds that religious expressions and behaviours are immediate, natural aspects of our human life, and not intellectual responses to the world. The *existentialist* conception holds that religious belief is closely related to profoundly meaningful experiences. The *nonsensicalist* conception holds that religious language, in grasping for the supernatural, exceeds the possibilities of sense for language; thus however valuable religious language may be, it fails, strictly speaking, to say something about the world or its intended subject.

By 'conception', Andrejč has in mind Wittgenstein's philosophical clarifications, principally as seen in his later philosophy but also evident in the early philosophy of the *Tractatus* and 'Lecture on Ethics'. A Wittgensteinian conception in Andrejč's accounting, drawing in part on the analysis of Oskari Kuusela, depicts an aspect of a phenomenon in order to elicit new insights about it even while remaining open to the combination of multiple perspectives. The idea here is that of a perspicuous representation: through creative and apt juxtaposition unseen aspects of a phenomenon become evident, especially a phenomenon concerning language.[8] For example, Wittgenstein first may refer to a description of a religious phenomenon, such as a claim involving theistic belief, as one may find in the 'Lectures on Religious Belief', and then respond in a way that questions the framing of the phenomenon instead of taking it for granted; this results in a move from considering the truth or falsity of the claim to contemplation of the possibilities of meaning for the claim.[9] Andrejč argues that different sources in Wittgenstein's corpus display different concerns and preoccupations in Wittgenstein's philosophical development and that the different conceptions across the remarks on religion evince his dynamic philosophical trajectory. Andrejč also reminds the reader that these conceptions are not unitary within texts; different conceptions may appear alongside one another.

Andrejč's identification of four conceptions of religion in Wittgenstein may be useful hermeneutically, and thus, I think it is a good idea to draw on them in order to see what is salient in Wittgenstein and might otherwise be missed. As with applying any hermeneutical scheme, there is a danger in over-interpreting these conceptions into Wittgenstein's writings, something Andrejč cautions against. Wittgenstein's remarks on religion tend to be reactive, and scholars sometimes can only infer or guess at what the stimulus might have been to produce his response; thus reconstructions of Wittgenstein's moments of clarification may be incomplete or imaginary. Andrejč advises that this list of conceptions should be seen as dynamic. In what follows, the conceptions that will be the primary focus of this chapter will be the 'grammaticalist', 'instinctivist' and 'existential' ones.

The 'Lectures on Religious Belief', published in 1966, is one of the earliest collections of Wittgenstein's remarks on religion to receive publication. In Yorick Smythies's notes from Wittgenstein's Lectures at Cambridge,[10] Wittgenstein is described memorably as observing a profound epistemic gap between one who believes in certain sorts of Christian doctrinal claims and another person who does not believe such claims. In portraying this profound epistemic gap between religious believer and non-believer, Wittgenstein appears to be setting up a stark divergence between a naturalistic epistemic standpoint and a religious one, a divergence registered by distinct 'grammar' of belief. Of course, while sometimes such profound gulfs may emerge, Wittgenstein's own treatment of religiosities across his corpus shows a more nuanced treatment of the relationship between religiosities and naturalism. Early in the first lecture, the text has Wittgenstein remark: 'Suppose that someone believed in the Last Judgment, and I don't, does this mean that I believe the opposite to him, just that there won't be such a thing? I would say: "not at all, or not always."'[11] What is perhaps most puzzling and provocative about this passage is the refusal to recognize this epistemic difference as being characterized by contradiction.[12] In Andrejč's terminology, this gap would reflect a grammaticalist conception of religiosity; the epistemic difference has to do with conceptual frames, not with acceptance or rejection of particular semantic content. The lectures go on to explore dynamics relating to different kinds of beliefs and the irrelevance of reason, evidence and contradiction to epistemic gaps of this and similar sorts.

In the 'Remarks on Frazer's *Golden Bough*', first appearing in 1967, Wittgenstein offers a series of engagements with James Frazer's monumental work on mythology and religion, *The Golden Bough*, which first appeared in 1890 in two volumes, expanded to twelve volumes by 1915, and abridged to a single volume in 1922.[13] Drawing on Andrejč's work, the Remarks on Frazer would be a key source for an 'instinctivist' conception of religiosity.[14] While Frazer's work was very well known in its day, it has been completely displaced in anthropology.[15] Perhaps with some irony, then, Wittgenstein's remarks on Frazer continue to have an influence in philosophy of religion and religious studies. The text that has been published of Wittgenstein's remarks was composed on multiple occasions in the 1930s and 1940s.[16] As with his other writings, Wittgenstein's remarks do not constitute a systematic view on the topic of so-called 'primitive' human culture and religion; rather, they show responses to what Frazer has written. Wittgenstein's reactions are clearly relevant to a number of topics in philosophy of religion and anthropology, especially as scholars have in recent decades taken a more historiographical stance with respect to research on the history of cultures and religions.

What stands out in the 'Remarks on Frazer's *Golden Bough*' is Wittgenstein's contention that religious and magical practices and beliefs are *natural* aspects of human life, that the having of rituals and ceremonies is part of ordinary human behaviour; these behaviours and beliefs are not based on some theory or the result of some proto-scientific, evidence-based reasoning about the world.[17] Because Wittgenstein does not systematically develop this naturalistic approach in philosophy, it is better to think of it as a sensibility. Arguably, this orientation in his thinking runs through several texts, especially in texts from the last period of his life.[18] While the 'Remarks on Frazer's *Golden Bough*' is a relatively early site for the appearance of this naturalistic sensibility in Wittgenstein, it is also developed in *Philosophical Investigations* and *On Certainty*. In the Remarks on Frazer, Wittgenstein writes of the power of fire on the imagination of the 'awakening mind' of human beings:

> I don't mean that just *fire* must make an impression on everyone. Fire no more than any other phenomenon, and one thing will impress this person and another that. For no phenomenon is in itself particularly mysterious, but any of them can become so to us, and the characteristic feature of the awakening mind of man is precisely the fact that a phenomenon comes to have meaning for him. One could almost say that man is a ceremonial animal. That is, no doubt, partly wrong and partly nonsensical, but there is also something right about it.[19]

This expression, 'the awakening mind of man' – or awakening 'human spirit' in Brian R. Clack's rendering[20] – appears another time in the Remarks, perhaps invoking the notion of a moment when biological humans developed a capacity for culture and religion within their natural history. This naturalistic picture of religiosity does not seem to be completely captured by an instinctivist conception of religiosity; instead this text also retains an existentialist sensibility, as rituals also reflect a human pursuit of wonder making out of experiences of the cycle of life and death.

Clack also points out that Wittgenstein of the Remarks invokes the idea of language containing a mythology: 'An entire mythology is stored within our language.'[21] While this remark arises in connection with criticism of Frazer, it conveys the idea, as Clack puts it,

> What Wittgenstein appears to be saying is that we have deposited in our language (as fossils are deposited in rock strata) the remnants of a mythology, the beliefs of previous times. When we speak of 'fate taking a hand', or of fearing 'the wrath of the gods', we bring out of storage the mythology of our ancestors. All of this is entirely consonant with Wittgenstein's overall view of language, which he likened to a city.[22]

This sensibility in Wittgenstein is difficult to name; one possible name for it would be the 'worldview' (*Weltanschauung*) sensibility. As Clack observes, Wittgenstein is drawing on insights of Oswald Spengler's view of the historical development of a culture, and the ties that remain between contemporary, earlier and ancient points in the development of a social group. While perhaps similar to and perhaps overlapping with Andrejč's grammaticalist conception of religiosity, the idea of a world view also draws to the foreground the historical and cultural background framing an expression – such as of religiosity – which is absent in the idea of grammar.

In moments like these, one might find in the Remarks a source for a *sort* of naturalized approach to the study of religiosities in so far as religious phenomena are presented as ordinary human activities, rather than aberrations of human cognition or relics of earlier uneducated cultures. In response to Frazer's purported explanation of the killing of the priest-king of Nemi, Wittgenstein writes:

> The religious actions, or the religious life, of the priest-king are no different in kind from any genuinely religious action of today, for example, a confession of sins. This, too, admits of being '*explained*' and not explained. . . . The same savage, who stabs the picture of his enemy apparently in order to kill him, really builds his hut out of wood and carves his arrow skillfully and not in effigy.[23]

First, Wittgenstein is putting the culturally mainstream practices of contemporary English Christianity on a par with the so-called 'primitive' forms of religiosity analysed in Frazer's early anthropological work. This suggests a sense of equity when it comes to human religious and cultural practices. Wittgensteinians countering forms of cultural chauvinism may find inspiration in moments like this. Second, Wittgenstein is challenging the uncritical linking of religious beliefs with truth-oriented belief-forming practices. Whatever the best way to understand ways of life we commonly call 'religious' may be – something we only get clues about in the Remarks and elsewhere in Wittgenstein's corpus – religious beliefs are *poorly* understood as folk scientific explanations of natural phenomena. This is one reason why Wittgenstein tends to denounce scientism. It mistakes the nature of the phenomena naturalists seek to explain. While Wittgensteinian moves like this, especially when later employed systematically by those influenced by him, are sometimes criticized for being 'protective strategies',[24] Wittgenstein's concern is with curbing the smug certainties that can blind researchers from properly grasping their intended objects of study. Generally speaking, Wittgenstein anticipates that human cultures tend to have accurate and pragmatic understandings of the world. Ways of life we call 'religious' exist alongside or within ways of life that adeptly and accurately grasp their worlds. Of course, epistemic gaps sometimes occur within religions and cultures, or else there would never be change over time. Nevertheless, intercultural epistemic gaps are likely to appear greater than intra-cultural gaps. It is when discrete beliefs or practices of a way of life are juxtaposed with elements of a way of life of people historically or culturally distant that a gap emerges.

In both the Lectures and the Remarks, a normativity about religiosities emerges, again, a feature that Andrejč highlights alongside the more familiar descriptivist themes in Wittgensteinian philosophy of religion. While Wittgenstein is loath to ridicule what he takes to be sincere religiosity, he sometimes criticizes forms of religiosity that embrace cosmological theorizing or that make other sorts of metaphysical claims (e.g. as are sometimes found in connection with Roman Catholicism). What is common amongst these and other references to religion, as found, for example, in *Culture and Value* or in 'Movements of Thought' is a valuing of sincere forms of religiosity over against insincere forms. But this religious sincerity runs quite deep in Wittgenstein's estimation; it is not something easily exchanged. One can see this attitude on display in his perplexity towards what he takes to be the 'stupidity' and 'monstrosity' of anyone ever converting between Protestantism and Catholicism.[25] Perhaps Wittgenstein's concern is that such a change might be brought on by reasons: 'Something that can (now) make sense only as a tradition is changed like a conviction. It is as if someone wanted to exchange the burial rites of our country for those of another.'[26] The problem here is not the sincerity of the reasons but the change of tradition itself; somehow the possessing of a tradition is too personal, too close to oneself to be forsaken due to changing convictions. At a later point in *Culture and Value*, in a remark from 1948, Wittgenstein writes, 'Tradition is not something a man can learn; not a thread he can pick up when he feels like it; any more than a man can choose his own ancestors. Someone lacking a tradition who would like to have one is like a man unhappily in love.'[27] One wonders what Wittgenstein would make of today's diverse spiritual landscape, with various traditional as well as new religious movements from around the world present in many societies. Perhaps, the phenomenon of being 'spiritual but not religious' would not be happily welcomed by Wittgenstein. And yet, if one has not received a tradition, or if a received tradition does not speak meaningfully to a new generation's circumstances, then a seeking for that which is felt to be absent might well be itself an emerging way of life.

What this suggests is that Wittgenstein saw the developing and inheriting of traditions as completely ordinary aspects of human life, but elements that were very important for meaningful living. In Wittgenstein's view, once developed, religious traditions may be deeply adhered to and coexist with beliefs and practices of an eminently practical and veridical sort. Implicitly, it is also possible for religious beliefs and practices to tread onto the realm of philosophy and theory-making, but when they do so, they become in his view inauthentic, perhaps even 'monstrous'.

7.3 Naturalism and Wittgenstein

It might seem that naturalism and Wittgenstein are an unlikely pair. After all, Wittgenstein is famously critical of our historical era's obsession with science as the paradigm of knowledge.[28] Wittgenstein's critique of scientism is arguably a theme that runs throughout his corpus, but that is not to say that it is a single theme in his thought. Jonathan Beale and Ian James Kidd observe that numerous forms of anti-scientism appear across Wittgenstein's corpus: 'being scientistic about philosophy, for instance, might not be the same as being scientistic about art or the understanding of religious beliefs'.[29] Many would suppose that naturalism and science, at least, are closely related and thus that Wittgenstein's critique of scientism would likewise entail a critique of naturalism. Depending how we define our terms, of course, that supposition would likely be correct. In the same volume, Benedict Smith writes:

> Naturalism is a view about the relation between science and philosophy, one that privileges the metaphysical and epistemological status of science. As such naturalism is usually understood as a view that regards philosophy as constrained in various ways by scientific methods and results. . . . Thus scientism could be construed as an especially strong form of naturalism: the relevant constraint is not just that philosophical inquiry ought to proceed in light of natural science, but that philosophy must itself be part of science if it can contribute to the ways in which we investigate and come to enjoy genuine knowledge of the world.[30]

Smith's definitions of these terms, 'naturalism' and 'scientism', are fairly standard and rightly reflect the mutually supportive relationship between naturalism and science; in describing 'scientism' as 'an especially strong form of naturalism', Smith construes scientism as being on a spectrum of forms of naturalism. Not all forms of naturalism will entail scientism, of course, but colloquially, naturalism sometimes carries the connotations of scientism.[31] We can see this reductive form of naturalism on display in the work of New Atheist philosopher Daniel Dennett.

Dennett's robust scepticism about religions stems from his understanding of religions being fundamentally incompatible with naturalist metaphysical commitments. Dennett defines religion 'as social systems whose participants avow belief in a supernatural agent or agents whose approval is to be sought'.[32] Dennett observes that supernatural agents may or may not be *very* anthropomorphic, but in his view, avowal of belief in a supernatural agent of some sort is a necessary condition for something being a religion. Of course, philosophers and scholars of religious studies challenge Dennett's definition;[33] it is *limiting*, leaving out putative religions that lack belief in supernatural agents (e.g. atheistic forms of Buddhism) and *distorting*, privileging supernatural belief over against other phenomena that particular religions might prize (e.g. Daoist ritual activity at home or in temples). Moreover, it is indifferent to research concerning the shifting meanings of 'religion' in history. Nevertheless, it is concise and enables Dennett to grasp the target he wishes to address in his work: contemporary religious beliefs that run counter to science and thus that he thinks are potentially a social threat.[34] Earlier in *Breaking the Spell* he shows awareness of family resemblance approaches to defining religion,[35] but since his philosophical concern is epistemically errant and socially dangerous forms of religion, his definition is anchored to what he takes to be the core of dangerous religion: avowal of belief in a supernatural agent and logically subsequent performance of duties to please that agent. In this way, Dennett is a clear and eloquent exemplar of a sceptical naturalism that is uninterested in hermeneutical richness; instead, this naturalism strives to 'break the spell' of potentially dangerous forms of religion.

At its most basic, naturalism can be defined in terms of metaphysics or method. Metaphysical or ontological naturalism today would be the view that 'all spatiotemporal entities must be identical to or metaphysically constituted by physical entities. . . . [Naturalists] hold that there is nothing more to the mental, biological and social realms than arrangements of physical entities'.[36] Another view sometimes held by metaphysical naturalists is methodological naturalism, the view that the methods of natural science are the best, sometimes *only*, way to know the world. Methodological naturalism would also hold that philosophy ultimately aims at the same kind of knowledge as the sciences do, even if their fields of focus are different (with philosophy often playing a meta-level role with respect to scientific inquiry).

Not all naturalists about religion need to embrace some form of metaphysical or methodological naturalism as just described, but conventionally, those like Dennett do. A metaphysical naturalist would be one who rejects implicitly the existence of gods, spirits, souls and other entities that are not, at bottom, physical. A methodological naturalist, in this sense, would be one who thinks that whatever the constituents of the world may be, they may be known best through the methods of natural science. In these senses, it seems that Dennett would be a naturalist of both the metaphysical sort – he does not think religious beliefs are veridical – and the methodological sort – religious beliefs can be explained through scientific studies of evolutionarily acquired human belief-forming processes.

Discerning Wittgenstein's relationship to the forms of naturalism is not an easy question, especially as it bears on topics in the philosophy of religion. Kai Nielsen's critique of so-called 'Wittgensteinian fideism' in his 1967 paper of the same name was grounded in his sense that the first generation of Wittgensteinian philosophers of religion were unreasonably trying to immunize religious language from metaphysically naturalistic criticism.[37] Nielsen's idea is that religious language (e.g. reference to God) was held by Wittgensteinians to be made meaningful only by the language-games and forms of life in which such language was contextually situated. Insofar as the conditions for meaningfulness were satisfied, then a religious person could respond to naturalist critics of religion that they really did not understand the religious language they claimed to be criticizing. Nielsen argued on the contrary that taking seriously the point of view of participants in a religious discourse does not preclude potential criticism of that discourse as incoherent and irrational. A vexing debate ensued over the following decades between Nielsen and D. Z. Phillips, amongst others, not just over whether Wittgensteinian philosophy of religion really was fideistic but also over the very standards by which such a question could be judged and what was at stake in such a controversy. What is relevant for our purposes is Nielsen's contention that Wittgensteinian philosophers displayed a hostility towards naturalism in philosophy of religion, and Phillips's parallel contention that empiricism in philosophy of religion tended to distort that which it sought to understand and explain, that is, linguistic expressions of religious faith.[38] One would be forgiven for thinking that naturalism and Wittgensteinian approaches to philosophy were fundamentally at odds with one another.

While Wittgenstein was a keen critic of scientism, and thus implicitly a critic of strong versions of methodological naturalism, at times, he might well have accepted some form of non-reductive naturalism.[39] Something adjacent to methodological naturalism may be present in directives such as 'don't think but look!'.[40] Chon Tejedor observes that Wittgenstein's anti-scientism is not limited to concern over scientific practices encroaching on 'areas of our lives in which they do not belong';[41] Wittgenstein was also motivated to criticize scientistic attitudes in order to protect the integrity of science, and in his criticisms of Frazer's methods in the Remarks, no less.[42] Furthermore, in *Philosophical Investigations,* we see references to human 'natural history', especially the natural history of language (§23). Human natural history is relevant to the human form of life and the possibilities which derive from it. Annalisa Coliva sees in the Remarks on Frazer a trace of this appreciation for the common propensities of humanity, 'a common way of feeling and being impressed by the same types of events'.[43] In this way, Wittgenstein's philosophical methods and focus are *this worldly*, upon the circumstances here and now, or there and then, of human lives with language and social practice, and he is concerned with the use of methods best suited to seeing linguistic and social phenomena clearly.

Might we speak here of a 'naturalistic spirit', taking some inspiration from Cora Diamond's reference to 'the realistic spirit' in Wittgenstein?[44] Some scholars, such as Sorin Bangu, have already drawn such a parallel.[45] A relevant key quotation of Wittgenstein's for Diamond is, 'Not empiricism and yet realism in philosophy, that is the hardest thing.'[46] In this way, there is a naturalistic mood or sensibility present in Wittgenstein's philosophical methods. Wittgenstein is concerned with describing what people actually do when they *live* or *perform* their religiosities, and less concerned about explaining the *truth behind* particular religious claims or beliefs. By turning his attention to understanding the particularities of human behaviour rather than a causal chain explaining away the holding of theistic claims, Wittgenstein's philosophy has a naturalistic sensibility. In so doing, then his philosophy finds common cause with the sociological and historiographical turns that have taken place in religious studies in recent decades when it comes to the study of religious phenomena as well as the category of religion itself.

7.4 Interpreting religious phenomena

When assessing whether religion is a natural phenomenon, it is necessary to have a working idea of the nature of religion, or at least of the aspects of those things we call 'religions' one is going to consider philosophically. As one can see from numerous essays and books in the field of religious studies, the definability of 'religion' is currently a hot topic.[47] It has become common in recent times for religious studies scholars to argue that the term 'religion' does not refer to a singular and distinct phenomenon[48] or to argue that the historiography of the term reveals a biased orientation towards Christian forms of piety, thus potentially distorting features of other traditions when the term is applied *generally* to ways of life in different cultures. There may be a sort of consensus amongst scholars about 'religion' insofar as it is not taken to be a natural kind term, that it is a social construct pragmatically taken – when it is taken up – to refer to a wide variety of beliefs, practices, institutions and other phenomena. A concern expressed periodically by religious studies scholars is that naturalists and cognitive scientists of religion tend to be overly reductive in framing religiosity in a way that can be philosophically explained naturalistically.[49]

While Dennett himself is evidently aware of the difficulty involved in defining religion in *Breaking the Spell*, he nevertheless defines religions in terms of 'belief in a supernatural agent or agents whose approval is to be sought'.[50] For those who wish to study religions with hermeneutical richness, there is much that is problematic in

Dennett's definition. His definition misconstrues salient features of religiosity to which Wittgenstein would advise us to direct our attention.

Religion in a Wittgensteinian vein – especially, but not only, as found in the 'Remarks on Frazer's *Golden Bough*' – typically emphasizes not just beliefs but also practices, including the practices in which language is used. 'Prayer' is listed as an instance of a language-game early in *Philosophical Investigations* (§22), and in *On Certainty*, religious beliefs (such as belief in God or in the authority of the Bible) are listed as being amongst the sorts of beliefs that are built on trusting relationships, for example, between children and adults, rather than evidence (e.g. §§106–8). This is not to say that evidence plays *no role* in the development or updating of beliefs of a religious nature, but importantly, it would need to be the right kind of evidence, perhaps given to an epistemic agent by a person who is in the right sort of relationship with the agent (e.g. a parent, teacher or co-religionist). To be sure, Wittgenstein decries the appeal to reasons and evidence in the 'Lectures on Religious Belief', but in *On Certainty*, we also see Wittgenstein referring to 'grounds' playing a role in persuasion vis-à-vis an adult educating a child to believe in God (§107). Grounds may be given for or against a religious belief, such as in God or in the authority of the Bible, but whether or not those grounds will be compelling may depend upon a network of personal relationships and the way of life for which belief or disbelief may play a significant role.[51]

Overall, for Wittgenstein, religiosities have not to do with evidence so much as with intimate social relationships and regular rituals and practices. Important to both *Philosophical Investigations* and *On Certainty* but especially to the *Investigations* is the idea that languages are social phenomena; thus, any religiosities insofar as they make use of language would be also social phenomena. Here, of course, it is important to remember that Wittgenstein held that religious beliefs were orientations quite different from ordinary beliefs about the world (e.g. perceptual beliefs). In texts like the 'Lectures on Religious Belief' and *On Certainty*, religious beliefs are not mere propositional attitudes about supernatural agents but forms of commitment that ground a whole way of life. In works such as the 'Remarks on Frazer's *Golden Bough*', the 'Lectures on Religious Belief', and 'Movements of Thought', additional religious phenomena include the holding of certain sorts of belief (in God, in the saving power of Christ), virtues of faith, the sense of religion being a comprehensive way of life, the performance of ritual sacrifices and the giving of religious arguments. Even with that understanding, Wittgenstein grasped that religiosity was much, much more than religious belief, and furthermore, that religious beliefs were not necessarily the core of religiosity but just one aspect of it – a very important aspect to be sure, but still, just one – amongst others (e.g. the ritual and the social).

Embracing the idea that study of religion is the study of what people do in connection with their traditions, religious studies scholars today study a very diverse collection of phenomena alongside beliefs and ritual practices in order to grasp the particularity and comprehensiveness of a religious way of life. For example, Ninian Smart had a growing list of non-exclusive 'dimensions' to be singled out as being relevant to the study of religious worldviews. In *Dimensions of the Sacred* (1996), published near the end of Smart's life, the list had grown to eight or even nine dimensions: the ritual/practical, doctrinal/philosophical, mythic/narrative, ethical/legal, experiential/emotional/, social/institutional, material/artistic as well as the political and the economic. Bruce Lincoln lists four domains that any satisfactory conception of religion needs to include; Lincoln's list includes discourses, practices, communities and institutions. Lincoln includes discourse 'whose concerns transcend the human, temporal, and contingent, and that claims for itself a similarly transcendent status'.[52] The idea here is to study not so much decontextualized doctrines but the roles that doctrines play in the give-and-take of training in a way of life or the manner in which these doctrines are interpreted and reinterpreted by religious communities in different historical or social contexts. Tomoko Masuzawa, moreover, has explored the genealogy of 'religion' and 'world religion' over the last century and a half, identifying the rise and spread of the paradigm of religiosity from the West around the world.[53]

These religious studies scholars, in disparate ways, describe the phenomena of religiosities in ways that are consistent with the spirit of Wittgenstein's descriptive remarks on religion and his philosophical methods.[54] While these recent scholars go far beyond Wittgenstein, their approaches are compatible with Wittgensteinian sensibilities: the reluctance to reduce the ragged complexity of religions in Smart, the attention to the flow of discourses over against abstract doctrines in Lincoln, the awareness in Masuzawa of contexts of naming and classification of something as a religion and the sense that inattention to the variety of uses of a term can invite confusion. Smart integrates a study of devotional practices, doctrinal beliefs, sacred stories, practical values, emotional experiences, social institutions, material culture, and economic and political relationships into his picture of religiosity, while Lincoln grasps the categories of discourse – along with ritual and ethical practices, local communities and larger regulatory institutions – into his conception of religiosity. Masuzawa further observes the social and political conditions of the construction of the category of 'world religion' that scholars like Smart and Lincoln seek to adapt to make more comprehensive. What I take from all three is the sense that analyses of religiosities may well take into consideration a wide variety of phenomena, including the social and political phenomena of the categorization of ways of life as religious. Note that the phenomena identified by these scholars are naturalistic in the sense that they are publicly accessible to study and observation; with the possible exception of emotional experiences and their subjectivity (Smart) – which may in any case involve social aspects, if the private language remarks in *Philosophical Investigations* are any guide[55] – all other dimensions mentioned by Smart and all those mentioned by Lincoln and Masuzawa are publicly accessible. Together, they serve to remind scholars that naturalistic study of religions need not be reductive; it can be fully compatible with the offering of thick descriptions of religions or contexts of ascription of religion-status.

A reductive naturalist approach to studying religion may encourage conceptions of religiosity that distort the phenomena supposedly being investigated. Some relevant examples that challenge the interpretive adequacy of reductive naturalism emerge from a project I assign to students in my Philosophy of Religion course at the Chinese University of Hong Kong, Shenzhen; in this assignment, students are tasked with interviewing some of their classmates about matters relating to beliefs and practices that are associated with religions. One characteristic example is as follows: just before taking the university entrance exam (the *gaokao*), a high school student in China visits a local Buddhist temple, lights incense and bows marking the four directions, and then kneels before an image of the Buddha in prayer. The student performs these actions not necessarily because she believes certain truth claims about the Buddha's powerful abilities and will to act in the world to benefit those who show devotion with a good exam score, but a significant number of students may perform these or similar acts anyway. As one interviewee put it: 'it can't hurt'. In fact, those same students do believe that years of very hard work studying for the university entrance exam will ensure the best possible score; and that is surely why so many students work countless hours preparing for the exam. If asked, the students who visit the local temple would likely claim not to be Buddhists and perhaps would identify as atheists. Even so, a significant minority of university students who disavow belief in supernatural beings might still believe in karma or that biological death is not the end of personal identity or that ancestors still exist in some sense beyond their deaths.

What this suggests to me is that the spiritual or religious imagination of university students in China – amongst other places – is far more vibrant and diverse than labels such as 'Marxist atheist' would seem to suggest. Perhaps in most people's minds, philosophical and religious beliefs are not held in a systematic, logical way; people who identify as atheists may also perform religious rituals and harbour religious imaginations that go far beyond what metaphysical or methodological naturalism would suggest is consistent with atheism. Actual practices and beliefs having to do with religions may run counter to preconceptions about traditional practices and religiosity itself, such as with pragmatically motivated temple piety. Lastly, the category of 'religion' (*zongjiao*) itself may be linked most strongly with traditions perceived as foreign, such as Christianity and Islam, but not with longstanding Chinese traditions, such as folk religion, Buddhism,[56] Confucianism and Daoism.

7.5 Conclusion

Wittgenstein has a great deal to offer for approaching the study of religious or quasi-religious phenomena like these, and such study would have to do, above all, with not merely 'thinking' but 'looking'.[57] This non-reductive naturalistic depiction is part of the clarificatory work of philosophy. It is valuable to be reminded that religiosities take countlessly diverse forms across societies around the world and further that conceptions of religiosities take diverse forms across societies because one can lose sight of these differences when doing philosophy. The craving for a master theory that can grasp and explain all, or nearly all, forms of religiosity through, for example, the adaptive or by-productive functioning of a neural mechanism has obvious attraction, but following Wittgenstein and student accounts like those just mentioned, such broad theories just are not plausible as explanations in general. That is not to say that psychological explanations for certain kinds of supernatural beliefs and their effects on behaviour cannot be gained through naturalistic or cognitive theories of religion; I have no doubt that much can be learned from such studies. Yet, modest and narrow claims ought to be the order of the day for studying phenomena as varied and unpredictably diverse as religions.[58]

Notes

1 McCauley 2020, 112.
2 Dennett 2006, 17.
3 Nielsen 2005.
4 Visala 2011.
5 In line with his times, Wittgenstein was not critical in his use of 'religion'; he seems to have taken it as given that the term could be used to grasp transhistorical and transcultural phenomena. When discussing Wittgenstein's views, I engage with his mode of expression even though a contextually sensitive family resemblance approach to analysing religion is arguably more fitting with his overall philosophy. See Carroll 2019 and section 3 for more on this.
6 Andrejč 2016, 19.
7 Andrejč 2016, 50.
8 Kuusela 2014, 73.
9 Phillips 2001, 5.
10 Volker Munz and Bernhard Ritter, in their introductory remarks on to 'Lectures on Similarity' in *Wittgenstein's Whewell Court Lectures: Cambridge 1938-41*, argue that Cyril Barrett's 1966 edition of the 'Lectures on Religious Belief' does not exactly correspond to available manuscript lecture notes. The 'Lectures on Religious Belief' may well come from multiple courses Wittgenstein gave between 1938 and 1940 (Munz and Ritter 2017, 86).
11 Wittgenstein 1966, 53.
12 c.f. Diamond 2005, 102.
13 Wittgenstein 1993, 115.
14 Andrejč 2016, 48.
15 Das 2020, 247.
16 See Albinus, Rothhaupt and Seery 2016 for more on the history of this document and the history of its publication.
17 Clack 2001, 22.
18 Plant 2011, 184.
19 Wittgenstein 1993, 129.
20 Clack 1999, 182.
21 Wittgenstein 1993, 133.
22 Clack 1999, 167–8.
23 Wittgenstein 1993, 123–5.
24 Carroll 2014, 93.
25 Wittgenstein 2003, 77.
26 Wittgenstein 2003.
27 Wittgenstein 1998a, 86e.
28 Beale and Kidd 2017, 1.
29 Beale and Kidd 2017, 2.

30 Smith 2017, 210.
31 Here we might also compare Kai Nielsen's description of naturalism in his *Naturalism and Religion*: 'Naturalism denies that there are any spiritual or supernatural realities. There are, that is, no purely mental substances and there are no supernatural realities transcendent to the world or at least we have no sound grounds for believing that there are such realities or perhaps even for believing that there could be such realities. Naturalism sometimes has been reductionistic (claiming that all talk of the mental can be translated into purely physicalist terms) or scientistic (claiming that what science cannot tell us humankind cannot know.)' (Nielsen 2001, 29).
32 Dennett 2006, 9.
33 Chaves 2010, 9; Crane 2017, 11–12.
34 2006, 14.
35 2006, 8.
36 Papineau 2020.
37 Nielsen 1967, 209.
38 Carroll 2010, 112 and 2014, 151.
39 Smith 2017.
40 Wittgenstein 2009, §66; Bangu 2018.
41 Tejedor 2017, 7.
42 Tejedor 2017, 22.
43 Coliva 2017, 46.
44 Diamond 1995.
45 Bangu 2018.
46 Wittgenstein 1978, VI–23.
47 Smart 1996; Lincoln 2003; Masuzawa 2005; Schilbrack 2014.
48 See, for example, Smart 1996.
49 Smith 2009, Saler 2010.
50 Dennett 2006, 9.
51 Carroll 2017, 760–1.
52 Lincoln 2003, 5.
53 Masuzawa 2005.
54 Carroll 2019, 54–5.
55 Carroll 2014, 157.
56 Even though Buddhism is a foreign religion, it has been present in China in some capacity for roughly 2,000 years. While it faces some challenges in contemporary Chinese society, perception of its sinification runs deep.
57 Wittgenstein 2009, 31.
58 I am grateful to detailed comments and helpful suggestions from Robert Vinten and Guy Axtell on previous versions of this chapter. I am also thankful for comments and questions raised by virtual attendees of the second ERB workshop held online in 2020. Any errors belong to the author.

8

Natural thoughts and unnatural 'oughts'

Lessing, Wittgenstein and contemporary CSR

Guy Axtell

This, then, is the ugly broad ditch which I cannot get across, however often and however earnestly I have tried to make the leap.
– Gotthold Lessing, *On the Proof of the Spirit and of Power* (1777)

8.1 Introduction: Wittgenstein on Russell's sceptical ire and O'Hara's friendly fire

Wittgenstein's 'Lectures on Religious Belief' (LRB) provide a source for as yet unexplored connections to religious ideas as treated in Robert N. McCauley's book *Why Religion Is Natural and Science Is Not* (2013). Pascal Boyer, Scott Atran and McCauley are amongst the leading CSR scholars who point out that 'cognitively speaking it is religion that is natural and science that is largely unnatural'.[1] CSR's concern with universal cognitive architecture is a concern with *etiologically deep* sources of religious ideas, and not with what is *etiologically shallow*, in the sense of ideas acquired along with culture, or religion-specific theologies. While Wittgenstein's Lectures focus around the examples such as a person holding a biblically based Day of Judgement belief, I will try to show that the questions which Boyer, Cohen and McCauley ask, questions which elucidate the many tensions between natural thoughts and unnatural oughts, are also quite directly engaged by Wittgenstein. They are engaged not only when he is expressing what Gorazd Andrejč refers to as an *instinctivist model* but also in the many ways in which a *grammaticalist*, or *existentialist* or *mystical* models. Each of these four accounts help Wittgenstein present alternatives to approaching religious ideas as do those who present science and faith as necessarily in conflict.

The radical religious fideism which Wittgenstein contemplates in LRB through his example of one Father O'Hara, as well as the sceptical evidentialism he attributes to Bertrand Russell, are ways of approaching religious ideas that Wittgenstein thinks of as poorly equipped to comprehend religious ideas. The sceptical evidentialism of Russell and members of the Vienna Circle, as well as the apologetic project of Father O'Hara, who takes his biblical Day of Judgement belief as one for which he can (and should) give compelling epistemic reasons, prompts Wittgenstein in one of his personal correspondences to quip that 'Russell and the parsons between them have done infinite harm, infinite harm'.[2]

These sentiments are often attributed to Wittgenstein's distaste for Conflict models of the relationship between religion and science, or faith and reason and his constant attempts to articulate an alternative Independence model. This is one helpful way to look at it: Wittgenstein's comment putting Russell and O'Hara in the same basket anticipates Ian Barbour's (1990) description of a Conflict model, since Barbour takes Conflict as having not one but two sources, one secular and one religious. The secular form comes either with a scientific materialism (confusing science and metaphysics) or with a stern religious evidentialist ethics of belief (confusing faith with empirically grounded belief, and psychographic contrariety with propositional disagreements). The religious source for Conflict, whatever its other motivations, which Barbour describes simply as 'biblical literalism', more broadly involves naïve realism about religious language or absolutistic claims about special religious authority or special access to truth or moral law within the *home* religion.

While the promise of independence is that contrariety in linguistic, ritualistic and even theological practices need not be thought of as contradictions in belief: contrariety is not conflated with contradiction, and science and faith are not brought into a relationship where they conflict. Yet as Bob Plant points out, Wittgenstein's comments sometimes stray from descriptions of concept use, and reflect his own conception of faith and its differences with certain others: 'In [LRB], Wittgenstein clearly has a specific conception of what constitutes *genuine* religious belief. As such, his remarks are often (albeit covertly) normative rather than merely descriptive'.[3]

So, a way of thinking about theologically correct faith such as Wittgenstein attributes to Father O'Hara is suggested by Wittgenstein, as we will see, to be equally as 'unnatural' in its own way; people can still err or make 'blunders' in their language-game, and O'Hara 'cheats himself', and becomes 'ludicrous' in Wittgenstein's view of him.[4] But there is something deeper and more positive as well, which more careful attention to multiple tensions between what seems natural to think or to do, and normative demands of various kinds, can help us to unpack. In this chapter, we engage CSR research more directly by exploring tensions between people's 'maturationally natural' inclinations to adopt religious ideas and the 'unnatural' demands sometimes made upon people, either (a) by imposition of evidentialist norms over religious utterances or (b) by an agent's adherence to perceived demands of theological correctness and to a particularly demanding conception of genuine or authentic faith. Wittgenstein's discussions of Russell and O'Hara help also to articulate these two tensions, and to open up alternatives which to some degree assuage them. I argue that these tensions we will explore are not only illuminating of the more and less helpful models of the relationship between reason and faith but that they also provide new research questions which connect Wittgensteinian (or more broadly Continental) approaches to philosophy of religion, to contemporary CSR.[5]

Gorazd Andrejč, Thomas Carroll and other Wittgenstein scholars think it is best to characterize Wittgenstein's work as expressing not just Independence but several different more specific conceptions. Andrejč for example articulates the *grammaticalist* conception for which the later Wittgenstein is best known but also finds *instinctivist*, *existentialist* and *non-sensicalist* (or mystical) conceptions expressed throughout his writings. As a result of this plurality in his thought, we will proceed according to Andrejč and Carroll's advice that each conception has its own resources, and it is hazardous to interpret Wittgenstein's reflections as committing him only to a single conception.[6]

8.2 Maturationally natural ideas and the practised natural

CSR seeks explanations of cultural phenomena in terms of acquisition, representation and transmission involving cognitive capacities. These evolutionary and cognitive approaches display less interest in study of cultural differences, but need not exclude them. CSR researchers study what they call the content and context biases that play an important role in the acceptance and transmission of religious beliefs; but these biases are present in other domains as well. For the CSR scholars, human beings have specific biases that make them susceptible to acceptance and transmission of religious beliefs. Like philosophy of religion, there are both secular and religious CSR researchers, and they sometimes debate whether evolutionary explanations are 'reductionist', or compatible with religious realism. McCauley and other naturalists typically steer clear of reductionism, but insist that philosophers of religion, hyphenated or not, resist 'special pleading', and the methodological unnaturalness 'of insisting (1) that religion and religious experience, in particular, are unique and, therefore, (2) that religion requires special methods of study'.[7]

McCauley points out how, '[A]t least on some fronts science does not come at all naturally to humans'.[8] So normative goodness and badness do not easily line up with or against the maturationally natural/unnatural distinction. 'Oughts' or normative prescriptions can come from different sources, and these prescriptions coming from on high may seem natural or unnatural to different people in different ways. Still, these distinctions can help us see the functions of these prescriptions, and perhaps even to decide which 'oughts' to follow, when they collide. Maturationally natural ideas may be appealing, adaptive and easily transmissible. The naturalness of religious ideas in itself neither supports nor debunks them. But it also makes for the presumption that religious domain is no *exception* to the role of bias in cultural selection. Natural ideas can facilitate some things and impede others so that there is no clear evaluation of them as a class. An agent who ignores 'oughts' as unnatural when they have the force of inductive patterns to support them, exposes their professed beliefs to serious aetiological challenges. For violation of inductive normativity is always epistemically relevant, and is in all *other* domains clear grounds for censure.

McCauley describes CSR research as multidisciplinary and argues that it is best characterized as an explanatory pluralism. But he points out that numbering amongst the considerations that render scientific explanations important are 'their abilities to highlight patterns and to describe the mechanisms that produce them', and 'their coherence with what else is known and, especially, with other scientific knowledge'.

There are multiple research programmes in contemporary CSR, some more cognitivist than others. So, we should not overgeneralize about CSR any more than about Wittgenstein's thought reflecting only a single conception of religious language. As self-described 'cognitivists', McCauley and Cohen (2010) argue 'that the cross-cultural recurrence and historical persistence of religion is attributable to the cognitive naturalness of religious ideas, i.e., attributable to the readiness, the ease, and the speed with which human minds acquire and process popular religious representations'.[9]

At first glance, Wittgenstein's discussion of religious and everyday belief in his *Lectures on Religious Belief* would seem difficult to interpret in McCauley and Cohen's cognitivist terms, if not counterpoint to it.[10] For one thing, their approach leads to a focal interest in 'popular' religious representations, in contrast to interest in the dynamics of adherence to an orthodoxy or to 'theologically correct representations'.[11] McCauley finds most interest in those who opt out from demands of theological correctness; Wittgenstein and Kierkegaard, with those who opt in. A strong theme in McCauley's book is that '[N]o matter how meticulously theologians articulate doctrines or how strenuously ecclesiastical authorities police orthodoxy, human minds will regularly follow paths and introduce variations that more closely harmonize with their natural cognitive prejudices'.[12] But the primary example of religious belief Wittgenstein uses, the Day of Judgement teaching, is drawn from purported revelation within the Abrahamic family of narrative testimonial faith traditions, and Wittgenstein's special attention is on O'Hara's conception of faith and its peculiarly problematic features. McCauley focuses on the role of *content* biases, whereas it is arguably *context* biases that are most pertinent to Wittgenstein's studied examples. Study of context biases will include the early education through which a model of faith, norms of theological correctness and particular beliefs of a theological sort are typically acquired.[13]

Another reason for the apparent misfit is that there are other research programmes in CSR besides those of the cognitivists, and a more Humean approach like Atran's might seem a better match with Wittgenstein writings. Emotions play more than the 'minimal role' in the acquisition and transference of religious ideas that evolutionary scientists sometimes allege: 'Hume's proposals about our psychological biases and propensities receive a good deal of support from recent work in the social sciences'.[14] Atran and Norenzayan (2004) and others such as Collier (2014) for example argue that Boyer, McCauley and Cohen fail to provide a complete account of people's 'natural' belief in gods and spirits. While the cognitivists find emotion and value to be overemphasized in psychology of religion, Wittgenstein, like Lessing, James, Kierkegaard and others who emphasize religious experience and the insider's perspective, find emotion and value to be central to religious language and practice. There is more concern for phenomenology or insider's perspective, as when Wittgenstein writes, 'This message (the Gospels) is seized on by a human being believingly (i.e., lovingly): That is the certainty of this "taking-for-true", nothing else'.[15]

I will argue that despite these differences in focus, there are rich connections between Wittgenstein's reflections on religion, and contemporary cognitive science, especially where he and both branches of CSR are concerned with tensions between natural thoughts and unnatural oughts.[16] But it will remain important to us that affectively driven evaluations of total evidence, and fideistic models of faith which prescribe emotional affirmations of scriptural or theological teachings, are directly important to understanding these tensions. It is important for locating research questions at the intersection of CSR and Wittgensteinian thought, that we not neglect how religious adherents very often adopt a particular model or conception of authentic faith, along with more particular teachings.

In LRB, these concerns are expressed by Wittgenstein not just through his reflection on differences between passionately appropriated religious teachings and evidentially grounded historical statements but also between narrative testimony and everyday assertive testimony. Wittgenstein describes an adherent of a narrative testimonial tradition who comes into contact with objective patterns or general facts that might challenge their belief, but who then declares: 'No. There it will break down. . . . No induction.' Wittgenstein's comment on this response is: 'That is, as it were, part of the substance of the belief.'[17] In *Culture and Value*, he further writes, 'Belief is not any kind of occupation with the object of belief. Fear, however, longing, and hope, occupy themselves with their objects.'[18] These reflections again suggest that passions and values are strong determinants of religious affirmations, and that with any doxastic states induced by affect or emotion, the *basis* of one's belief is somehow partially constitutive/determinative of *the content* of one's belief.

We will return to develop Wittgenstein's reflections more fully later, but it may help to here clarify McCauley and his co-authors' thesis, in order to clarify my own thesis. McCauley and Graham (2020) understand maturationally natural systems as *untutored:* 'they are untaught and need not be taught. . . . The advent of these predilections of mind does not rely on training, teaching, cultural artifacts, or the special preparation of environments.'[19] This leads them to describe these systems as cognitive processes which are 'automatic, instantaneous, and unreflective'. To go further,

> The *maturational* naturalness of perception, cognition, or action contrasts with another form of implicit cognition, that is, with the *practiced* naturalness that various individuals may develop with some culturally specific materials. Practiced naturalness deals with the wide range of capacities that individuals exhibit, from basketmaking to bicycle riding, at which many can, *eventually* with lots of experience or practice, achieve an expertise.[20]

So, McCauley and Graham allow that 'Systematic reflection generates intellectual working space beyond that which our maturationally natural tendencies supply' and that 'matters of practiced naturalness are not just domain specific; they are also *culturally specific*'.[21]

On my view this re-invites the role of affect, and so also the study of religious fideism, and the role of a model or conception of faith for the religious themselves.[22] The practised natural, especially in relation to an inherited testimonial faith tradition, typically results in a religious adherent holding their theological beliefs together as a packet, rather than separately. This in turn has the consequence that agents often forget that the theologies which purport to explain these interconnections are secondary constructions, offering interpretation of the theological lessons of ancient sacred narratives. So, while there is much to be studied regarding popular religiosity, and popular religiosity may be the natural object of sciences rooted in evolutionary theory, I will be insisting again that cognitive psychologists not neglect what can be learned from studying religious fideism but instead study directly the impact which adoption of a specific conception of genuine faith or theological correctness has for situated religious adherents. My thesis is that the 'bottom up' of CSR's approach and 'top down' concern with agents conforming to very particular and sometimes very demanding models of faith and of theological correctness are *both* needed to understand tensions between natural thoughts and unnatural oughts.

In order to highlight a productive intersection between Wittgensteinian scholarship and CSR, the next section focuses close attention on Wittgenstein's response to Lessing's distinction between 'two ways' that an adherent of an Abrahamic religion may take in regard to faith-based belief, with narrative scriptural testimony and the 'Day of Judgment' belief, in particular, as his main example. Here we will see how 'top down' concerns are illuminated by responses to Lessing's problem, the problem of the 'ugly, broad ditch'. The subsequent sections go on to argue that Wittgenstein's subtle and searching reflections on the two ways are: (1) reflective of some of the most significant differences between moderate religious fideism and problematically strong fideism; (2) connected thematically with central hypotheses pursued through experimental means in contemporary CSR; and (3) illuminated descriptively through the *psychology* of luck and risk, and philosophically through recognition of the comparative riskiness of models of faith that associated with religious enthusiasm, the riskiness stemming from their prescribing a (to use Kierkegaard's term) 'teleological suspension' of the moral, or epistemological and logical.

8.3 Facing the 'ugly, broad ditch': Lessing, Wittgenstein and the unnatural 'oughts' of O'Hara's biblicism

In LRB, Wittgenstein asks us to 'suppose somebody made this guidance for this life: believing in the Last Judgment'.[23] Wittgenstein reflects on the different grammars of a professed Day of Judgement believer, in conversation with another person who does not profess to have that belief. They are depicted as talking past one another, and their communication is poor, since the Day of Judgement believer is not making a claim that can be assimilated to everyday or empirical claims, nor even historical ones: 'These controversies look quite different from any normal controversies. Reasons look entirely different from normal reasons. They are, in a way, quite inconclusive.'[24] Imagining various ways that such a person might attempt to converse with various other persons who do not share the belief but are interested to understand it allows Wittgenstein to compare and contrast the prescriptions of faith with scientific questions of the objective 'grounding' of belief. He goes on to comment, 'In a religious discourse we use such expressions as: "I believe that so and so will happen," and use them differently to the way we use them in science.[25] . . . [T]here is this extraordinary use of the word "believe." 'One talks of believing and at the same time one doesn't use "believe" as one does ordinarily. You might say (in the normal use): "You only believe – oh well. . . ."

Here it is used entirely differently; on the other hand it is not used as we generally use the word "know."'[26]

Wittgenstein also places O'Hara in an imagined conversation with others who may assent to or confess what he does, but who do not understand their faith-based assent in the same way as O'Hara, who is keen to offer what he takes to be evidentially sufficient support. Using the biblical 'Day of Judgment' teaching as his main example, Wittgenstein thus tries to assuage tensions in O'Hara's religious apologetics by explaining how for adherents of a testimonial faith tradition, the affirmation of scriptural teaching is not like that of everyday beliefs, or like a prediction based upon past experience. Rather, it draws upon an 'entirely different kind of reasoning'. More carefully, Wittgenstein writes that for believers themselves, either beliefs of this sort:

(1) 'are not treated as historical, empirical, propositions'; or
(2) they are treated as 'historical facts that are different from a belief in ordinary historical facts.'[27]

Wittgenstein here identifies 'two ways' which Christian thinkers have taken to religious claims stemming from scriptures. I believe that Gotthold Lessing's famous discussion of the 'ugly, broad ditch' faced by adherents of testimonial faith traditions supplies a backdrop for many of Wittgenstein's reflections in LRB, including this description of the two ways to treat narrative teachings drawn from scripture.[28] Since we begin with Lessing's problem and its image of a ditch or waterway, I will refer to those who ascribe to the first of these two ways as Left-Bankers, and those who ascribe to the second as Right-Bankers.

The literature on the ugly, broad ditch, or what we will simply term Lessing's problem, recognizes three distinct aspects or 'gaps', as Lessing first developed them. There is a different but related 'unnaturalness' of these temporal, metaphysical and existential gaps,[29] which frustrate self-attribution of religious goods in the manner of the second and more radical way. This is the way that O'Hara represents when Wittgenstein introduces him as 'one of those people who make it a question of science'.[30] My use of Left- and Right-Bankers is not intended to mirror Left and Right (wing) Wittgensteinians as they have been described by Bernard Williams (1992; 2005) or others, though there clearly is some strong overlap.[31] The 'two ways' seem to involve very different ways of setting or not setting religious beliefs apart from other domains. This passage might be used to explain a dividing point between self-described Wittgensteinians, insofar as liberal theologians from testimonial faith traditions tend to take this first option while conservative thought favours the second. Wittgenstein does not clearly promote one or the other of these two options; however, independence fits most comfortably with the first way, and Right-Bankers are in my view *disguised* Conflict theorists, even while (and by) explicitly embracing this contingent/necessary truth conflation as a core apologetic strategy. The reduction dictates that science or reason can never really challenge religious dogma, and it arguably treats reason only as a handmaid of theology or accepted orthodoxy or scripture. Wittgenstein's reflections on the two ways are informed by his close reading of Gotthold Lessing and Søren Kierkegaard, and their earlier struggles with the 'ugly broad ditch' problem that attends prescriptively fideistic conceptions of faith prevalent in the Abrahamic religions. A New Testament example of prescriptive fideism is Jn 20.29: 'Then Jesus told him, "Because you have seen me, you have believed; blessed are those who have not seen and yet have believed."'

While Wittgenstein does not unequivocally accept the first way and reject the second, he does explore in great depth how the second option – that scriptural narrative events, past or future, are historical facts, and evidence of them is objective but 'in a different way' from historical evidence – invites paradox.[32] O'Hara does *not* do what Wittgenstein prescribes: he treats it as another historical message and does not 'make a quite different place' for it than normal or everyday belief (CV, 37). One's faith-based belief being at odds with our everyday use of 'is certain' and even 'believes', doesn't necessarily undercut these self-attributions of belief and of undoubting surety or constancy. But he does illuminate the ditch, which is to say, the tensions with the use of 'believes', which correspond to Lessing's three 'gaps'. It also reveals much about the tensions within models of faith which are *prescriptively fideistic* by, for example, prescribing belief, prescribing certitude or constant undoubting. If applying objective reason to answer questions that are partly of the heart is radically counterintuitive or unnatural, as CSR describes science and critical thinking to appear to merely intuitive or maturationally natural thinking, then separating objective and epistemological 'reasons' and 'evidence' enough to refrain from self-attributing special access to truth is radically counterintuitive or unnatural to a religious enthusiast.

Mr Lewy, one of Wittgenstein's former students at Cambridge, and O'Hara were both persons who Wittgenstein actually knew, but in LRB he is merely using them to illustrate different kinds of thinkers. O'Hara, unfortunately, seems to be a concrete example of someone who, instead of recognizing different uses of language, asserts his Day of Judgement belief in such a way as to make it look either like religious authoritarianism (biblicism) or like the worst connotation of fideism: belief *grounded* on weak evidence. He does not do justice to the *prescribed* under-determination of faith by evidence; nor does he or Lewy provide a very functional response to the serious problems with which Lessing and Kierkegaard struggled. To refer again to the image of the 'ugly, broad ditch', O'Hara is a prime example of the second path, and of a Right-Banker to self-ascribe truths achieved through non-epistemic means. In *Problems of Religious Luck: Assessing the Limits of Reasonable Religious Disagreement* (2019), I described this as the urge to 'have one's ditch, and cross it, too'. Here I will elaborate these this point by contrasting O'Hara's response to the ditch problem as *akratic* or intellectually incontinent, with Left-Bank responses, which I see as *enkratic*, or continent (and more self-consistent) responses to Lessing's problem, and to the problems of religious contrariety.[33] Let us look at the first and second responses in more detail.[34]

8.3.1 Left-Bankers

Mill, James, Kierkegaard and Wittgenstein, each in their own subtle way, *deflates* the epistemic standing of faith-based states or acceptances. This expresses the first way, and a continence, or *enkratic* response to the Enlightenment challenge to religious enthusiasm: the self-attribution of belief but not knowledge and the self-denial of the pleasures of unconstrained religious absolutism. One may always wish to self-ascribe religion-specific knowledge and moral and epistemic superiority over unbelievers at the same time. But in resisting this 'natural' impulse to award oneself the religious 'prize', one instead remains true to what in the literature on exclusivist and pluralistic responses to religious contrariety, is referred to as the Intellectual Golden Rule (IGR). As Hick describes it, this is a rule imbibing me to dialectically grant to others a premise which we rely on ourselves: that our tradition's scriptural transmission, and instruction through elders, is sincere in its teachings and guidance.[35]

Left-Bankers acknowledge faith-based beliefs they acquire from culturally particular settings to be poor candidates for knowledge or objective justification. But this has nothing to do with faith's intellectual 'failures', but rather with the differences between faith-based commitments and inferences from evidence. The Left-Bankers refrain from indulging in knowledge-claims, or from asserting the historicity of events in the home religion's sacred narratives.[36] This self-limitation allows the faithful justification but not exclusive truth or religion-specific 'knowing'. To give a closer example, Benton (2006) argues that the triadic organization of Kierkegaard's response to Lessing (through the writings of his pseudonym Climacus) signals his awareness of the threefold nature of Lessing's problem. In this subjective response, 'Climacus maintains that "belief is not a knowledge but an act of freedom, an expression of the will". . . . Climacus here is pointing up the volitional, rather than cognitive, character of belief. Hence, "The conclusion of belief is no conclusion [*Slutning*] but a resolution [*Beslutning*], and thus doubt is excluded."'[37]

8.3.2 Right-Bankers

The second way seems to allow the Kierkegaardian move from faith to truth (Pojman 1986a) but does not well-acknowledge that it conflates narrative and assertive testimony, and potentially proliferates contrary, religion-specific 'truths' and 'knowledge'. The idea of religion-specific knowledge in the face of religious contrariety is a difficult concept, indeed. The problems of grounding a literal-historical interpretation of biblical teachings are severe. While Lessing says that he cannot honestly make the leap required to cross the ugly broad ditch, this problem is 'resolved' in the minds of many contemporary Protestant Christian philosophers simply by going *unacknowledged*. Lessing claimed that 'you can have true Christianity by living with love' and need not take the miraculous as historical. By contrast, William Lane Craig arguably epitomizes the Right-Bank response in an interview where he is asked directly about Lessing. Craig first responds that Christians are 'not dependent on historical proofs for knowledge of Christianity's truth', but then seems in his religious *epistemics* to negate the conceded logical and evidential under-determination, saying that 'through the immediate, inner witness of God's Holy Spirit every person can come to know the truth of the Gospel once he hears it. . . . So that's how I leap Lessing's ditch. Christian belief is confirmed by the historical evidence for those of us fortunate enough to be epistemically so situated as to be able to appraise it correctly; but Christian belief is not based on the historical evidence'.[38] So while the first way acknowledges a lack of criteria for self-evidence and the arbitrariness of intuitional faculties, the second way is akratic in asserting a unique phenomenology as a means to guarantee the Christian's religion-specific knowing.

8.3.3 Wittgenstein and O'Hara

Wittgenstein says of the Day of Judgement believer *not* that s/he's unreasonable, but rather just that s/he's 'not reasonable'. This *could* mean a blunder, but for the generic believer this would not necessarily justify censure, or as Wittgenstein puts it, 'rebuke'. It could mean instead that 'they don't use *reason* here'. That could just indicate that they are making a different place for it while remaining modest or enkratic; but right after this Wittgenstein, rather stunningly, turns back to Father O'Hara and turns *against* him in censure:

> I would definitely call O'Hara unreasonable. I would say, if this is religious belief, then it's all superstition. But I would ridicule it, not by saying it is based on insufficient evidence. I would say: here is a man who is cheating himself. You can say: this man is ridiculous because he believes, and bases it on weak reasons.[39]

This criticism results from O'Hara's failure to acknowledge what James calls the 'mood of faith' behind his evidentialist apologetic, and from his subsequent conflation between his subjective certainty and objective or epistemic certitude. O'Hara or someone like him assumes that to be reasonable in his faith he must offer an evidentialist apologetic, making this person a 'ludicrous' figure to him, Wittgenstein says. This surely is a term of censure or rebuke associated with making a 'blunder', even if 'a blunder in a particular [game or] system'.

O'Hara, a religious rationalist of sorts, holds the biblical 'Day of Judgment' belief as historical event, and his high credence in this narrative, as revelation or prophesy he would take to be supported by broader natural theology. Wittgenstein finds O'Hara's biblical evidentialism to be a misconstrual of faith; O'Hara does not recognize the mismatch between his faith-based belief (the assumption of biblical authority) and 'grounding' evidence for that belief. No doubt there are substantial differences between Catholic and Protestant conceptions of faith that could be coming into play, but both should lead adherents to acknowledge and embrace the 'mood of faith', as James would put it, in their affirmations. The more O'Hara or others appear blind to this themselves, the more apparent their confusion, in Wittgenstein's terms, symptomatic and criterial treatments of evidence.[40] One can of course have knowledge of a narrative, or of what a particular theory claims or entails, without thinking yourself well situated to have a settled view about whether the personages and events within the narrative are (or are even intended to be) historically accurate, or whether the theory is 'true' in some full-blown sense. The second way, in which a person self-attributes religious truth and knowledge, may seem more 'natural', yet far more epistemically risky than the other, which does not self-ascribe such full-blown knowledge.

Wittgenstein reflects on some epistemological implications of Lessing's problem:

> We don't talk about hypothesis, or about high probability. Nor about knowing. In a religious discourse we use such expressions as: 'I believe that so and so will happen,' and use them differently to the way in which we use them in science. Although the temptation is great to think we do. Because we do talk of evidence, and do talk of evidence by experience. (LRB, 55-56)

The expressions 'I believe that' and 'I believe in' reflect 'an entirely different kind of reasoning', which his imagined interlocutor should acknowledge, even if O'Hara himself does not. The interlocutor wonders what is meant, and sees contrariety if he does not similarly hold the Day of Judgement belief; but O'Hara supposes an undue clarity, in order to reduce the contrariety to propositional *contradiction*.[41]

One 'naturally' wants to self-privilege: to stand on the far bank, and not the near, Left-Bank of the ditch. One wants to suspend inductive norms, since one would otherwise have to deny that similarly situated inquirers cannot obtain the same epistemic or alethic goods in their own, contrary testimonial tradition. One wants to suspend universal aspects of reason, and yet not be challenged in one's faith-based beliefs. Yet counter-inductive inference, and historical belief on non-historical grounds, are equally 'unnatural'. The 'oughts' issued by prescriptively fideistic conceptions of faith appear to be both unfulfillable and unnatural. The criterial usage of language that goes best with the first of these two ways may produce independence and hence insulation much less paradoxically than the second way, where symptomatic (inductive or objective) reasoning is taken as applicable to biblical events. An apologist for the second of Wittgenstein's two ways demands explaining these 'facts' which are 'different' than how facts are understood in any other domain. Such apologetics are associated with insider/outsider binaries, and counter-inductive thinking to maintain them.

It is perhaps common to suppose that if I believe something, and if believing that *x* entails thinking that *x* is true, then my believing *x* must also entail that I *think* I know *x*, and that absent any undefeated defeater, I *do* know *x*. It is common, in other words, to assume that self-ascribing a belief *x*, is to ascribe to myself a knowledge-claim. But is it best to draw the major line between believing or withholding 'belief' where all other epistemic goods come in tow with belief? Or is it best to say one can self-ascribe religious beliefs, while yet withholding self-attribution of knowledge and understanding?

Right-Bankers typically hold conceptions of faith incompatible with allowing that doubt is a normal feature of religious life. To this extent, it is no less an 'ideal theory' than those that evidentialist maxims imbibe, that it is always wrong to believe anything on logically insufficient evidence. Indeed, the debate over the appropriateness of ideal theory for the normative tasks of epistemic and ethical evaluation is a debate caught up in assumptions which sometimes strike us as natural, and other times as quite unnatural. Why should this be less so when we include the normative or prescriptive posits of a model of authentic, or again of theologically orthodox, religious faith? Theologies, and especially accounts of divine providence and salvific value (soteriology), utilize ideal assumptions in describing faith, and in narratives featuring exemplars of faith.

Theological evaluation, that is, evaluation on the basis of ideal religious virtues, and character types who exemplify faith or one of its contraries, makes prescriptions that sometimes strike us as natural, and other times quite unnatural. Philosophical concerns about teleological 'suspensions'[42] of the moral or the epistemic, where these are understood as normative and based upon human 'universals' is thus at the same time a concern about how adherence to one theological method/model or another might affect not only assumptions about theological value but moral and epistemic value as well.[43]

Wittgenstein strongly (and wisely) resists any over-easy characterization of differences between the Day of Judgement believer and his sceptical interlocutor as being characterized by logical or propositional contradiction. To generate a contradiction, it appears that the agent must *intend* their affirmation be taken in a specific-enough sense that the other could deny what is affirmed.[44] But it is at least easier to generate such a contradiction between single statements, and Wittgenstein realizes that theological systems are secondary constructions. He seems to regard O'Hara's evidentialism as a 'blunder' in some neutral (philosophical) sense; it is a blunder within the particular system or game since it is both fideistic and rationalistic at once. In cases where beliefs that *need* not rub up against each other nevertheless do, perhaps because a speaker is pushed to give clear propositional content to what she claims to believe, or claims is true, further issues arise of what constitutes 'grounds' and why lacking grounds is important or not.

Christian philosopher Dick-Martin Grube (2015; 2005) explains how bivalence

> implies a particular way of dealing with that which is genuinely different: It implies that, if position A is true and position B differs from A, B *must* be false. [Yet] . . . B's falsity is not affirmed after careful scrutiny by default, viz., simply by virtue of the fact that A is held to be true. Under bivalent parameters, there is no other choice than to consider B to be false. Since only A *or* B can be true but not both, B must be false if A is true [so that] bivalence implies an *equation between difference and falsity*.[45]

Especially so when contraries are multiple, and represent systems with explanatory ambitions, inductive evidence begins to mount that the presumed contradictions are not so truth-apt as the propositional language assumes, and philosophers of religion and CSR should actively investigate the reduction of contrariety to contradiction. As psychologically driven inferences, biased-closure inferences reduce contrariety to simple contradiction, and the central theological or religious claims of the home religion as true or false, simpliciter. This type of formally fallacious inference is commonplace, insofar as the study of social biases related to belief polarization reveals it. But it is nurtured by religious apologetics which pretend to the logical point that 'there is at most one correct theological system', which in practice turns siblings into rivals in the Abrahamic family of testimonial faith traditions, over the one true religion, or most authentic or authoritative church, or set of scriptures.[46]

These may be other reasons why Wittgenstein surprises his readers in writing that he 'would definitely call O'Hara unreasonable'. O'Hara's stance suggests to me a want for recognition of 'the ugly, broad ditch', reflection on which Lessing, Kant, Kierkegaard, Wittgenstein and others provide such illuminating reflections. The Right-Bank response seems to raise the ante in various respects, and while a sufficient marker of fundamentalist religious orientation, it is associated by Wittgenstein with various ills or potential blunders. More contentious, it seems to make comparative philosophy of religion, and ecumenical efforts amongst theologians pointless, prescribing instead something similar to Luther's contention against Erasmus, that there is no place for a sceptical, but only for an assertive Christianity. The second point takes on positive apologetic projects which Wittgenstein and many others have thought are both philosophically confused or 'blundering', inimical to genuine faith to the extent that they deny the relevance of risky strategies of belief formation and maintenance to epistemological norms, and to the ethics of belief. Owning up to risk, and thus to our fallibility and doxastic responsibility, will be important in the remaining sections, as we consider further the adequacy (philosophically and also theologically) of various responses to Lessing's problem.

8.4 A philosophical refrain

We started by asking how Wittgenstein's provocative, even if somewhat inflated, 'infinite harm' statement, one which lumps religious and sceptical rationalists together in seemingly unnatural ways, can help us understand developments in Continental, post-Wittgensteinian philosophy of religion. We asked how Wittgenstein's critical stance towards them, and discussion of assumptions they share, might be of renewed interest to philosophers of religion (hyphenated or not), and to longstanding debates over science and religion. We used Wittgenstein's reflections in PRL and elsewhere, especially where they plausibly show a direct concern with what we called Lessing's problem, as a further sounding board for answering these questions, and for drawing out deep tensions between seemingly natural thoughts and seemingly unnatural oughts. And we connected and elaborated these tensions, and Wittgenstein's own interest in them, directly to some questions centrally asked in contemporary CSR. The questions of most pertinence to us were questions concerning how to understand tensions between our maturationally natural leanings with inductive norms and scientific reasoning on the one hand, and with the sometimes radically counterintuitive prescriptions of particular, often historically prevalent conception of authentic religious faith, on the other.

So, where does our exploration of this rich intersection between Wittgenstein and contemporary CSR leave us? It is of course presumptuous to say just what 'oughts' to follow, over even which oughts seem most natural and unnatural to different persons. But our point is not that unnatural oughts are good or bad, simpliciter; rather, it is that there are various thoughts about these pairs – natural and unnatural, good and bad – and that their intersections help us to ask pointed questions about how one can and should respond to them in different practices or domains of discourse. One thing I think we discovered is that the purposeful diminishment (from the absolutist's perspective) of the epistemic status of faith-based belief is neither incidental nor accidental to the trio of Kierkegaard, James and Wittgenstein. It is instead, as I take Wittgenstein to be showing us, coextensive both with the special value accorded to faith *by* the faithful and in another way with his working definition of philosophy and philosophical method as a 'battle against bewitchment of intelligence by means of language'.[47] Moreover, this epistemic diminishment is not taken as inimical to the value of faith ventures, since personal existential risk and genuine faith are fused for each of these writers.

Wittgenstein's reflections on symptomatic reasoning help to explain the strong *conceptual* connection between inductive strength/cogency and positive epistemic status, or candidacy for 'knowledge'. There are close conceptual connections between epistemic risk and violation of inductive norms, on the one hand, and between unsafe or risky aetiology of belief and low epistemic standing, on the other. Thus, I count Wittgenstein along with James and others as articulating mainly a Left-Bank perspective, reminding us that self-attributions of religious knowledge needlessly beget Conflict, and that genuine faith does not require such self-attributions or the apologetic projects they give rise to (in O'Hara, for example).

From the agent's perspective, the religious ideas one is immersed in are likely to have much personal appeal, but the *theological constraints* carried by the particular conception of faith one adopts can be quite demanding upon them. While McCauley tends to emphasize the contrast of religious ideas as natural, and scientific reasoning as often counterintuitive, he does also acknowledge the radical counterintuitiveness of some of the demands of 'theological correctness'. One may well recognize that one's primary use of religious concepts is 'criterial', not 'symptomatic' – not 'grounded' in the way that inference from evidence is, and that 'knowledge' is in the everyday and scientific sense. But some of the dictates of one's adopted model of faith are likely to appear quite unnatural. This will be so whether salvation is presented either in a radically deterministic/fatalist way or else in a radically voluntarist way, where full and unchanging belief is prescribed as a condition of correct or genuine faith.

But arguably, the 'ditch' problem, and Wittgenstein's reflections on it, exposes the unhappy mix of voluntarism and involuntarism in those who place themselves on the far side of the ditch, with inerrant or 'historical' belief. This problem makes all too evident the Right-Banker's bad faith with respect to their lack of acknowledgement of the epistemic risk they are taking. Voluntaristic assumptions are readily evident in divine command narratives, and in confessionalisms, evangelisms and biblicisms. O'Hara's apologetic is one where 'we risk pretty little', but presenting the matter so would be disingenuous, or the result of multiple logical blunders. Radical fideism often manifests as believing *on faith*, but in such manner that belief is treated as an *achievement* concept with respect to correct adoption of the theological orthodoxy of the true religion or sect. So, for both logical and psychological reasons, a great deal of unnaturalness attends authentic faith portrayed as involving certitude about matters of creed. Prescribed 'certitudes' of doctrinal voluntarisms exhibit logical gaps; they also invite cognitive dissonance on the part of the agent they become normative for – ironically even as faith is portrayed as permanently closing the gap between subjective and objective truth, or between existential and metaphysical concerns, by sheer goodness of will on the part of the confessing or 'believing' agent.

The special problem with a Right-Bank response to Lessing's problem is that it makes religious or theological discourse/practice an exception to how we look at all other truth claims, and to make this exception is, for any neutral inquirer, a most unnatural ought. Systems of theology often make radically counterintuitive prescriptions, and the 'historical' religions perhaps more so than most others, because of the fideistic models of faith that they incorporate. When a person is presented evidence of a pattern of which they are part, and this person says 'no induction', or in other words, 'the pattern stops here', why should we not take their response as important further evidence that they *do* fit the pattern which they deny? Counter-inductive thinking and the epistemic risk which attends it, in other words, is fertile soil for serious aetiological challenges.

I want to suggest that while Wittgenstein is trying to find intelligibility in religious cognition and its expression, what he is saying is not so different either from what is maintained by theologians like Paul Tillich or William Sessions, or from humanists like Philip Kitcher. Paul Tillich on my view has a concern with Lessing's problem in mind when he writes,

> The affirmation that Jesus is the Christ is an act of faith and consequently of daring courage. It is not an arbitrary leap into darkness but a decision in which elements of immediate participation and therefore certitude is mixed with elements of strangeness and therefore incertitude and doubt. But doubt is not the opposite of faith; it is an element of faith. Therefore, there is no faith without risk.[48]

William Sessions in his study of different models of faith writes, 'Non-evidential firm belief is central to faith on this [Belief] model. . . . The belief is a matter of conviction, not certainty. . . . Evidential risk is therefore ineradicable for the belief model, and it prevents faith from pretending it is knowledge.'[49]

Wittgenstein often engages with the expectation that persons should always be in a position to provide reasons for their beliefs. The misguided nature of this expectation, as Queloz (2016) points out, was for Wittgenstein traceable 'to the tendency to think of chains of reasons on the model of chains of causes'.[50] Interestingly, Philip Kitcher, similarly takes beliefs to be legitimately self-ascribable in ways that do not commit one to also self-ascribe propositional knowledge or warrant/grounds. An explicitly secular philosopher, Kitcher yet shows himself to be quite sensitive to how this sort of point applies to religious epistemics. Indeed, it informs his response to Richard Dawkins and to sceptical evidentialists more generally:

> Asking after the grounds of contemporary religious belief, and embarrassing the believer by demonstrating that the processes that underlie it are unreliable tries to confine devout people in places where they do not belong. Lack of epistemically secure grounds can simply be conceded. . . . The separation can be made in either of two distinct ways. . . . One can take faith to be a legitimate mode of grounding belief, even though the doctrines accepted do not count as items of knowledge. [Or] one can think of faith as a form of commitment, not expressed in beliefs at all.[51]

Sessions provides a description of moderate religious fideism, and Kitcher's matching two options allow for the naturalness of religious inclinations, and for moderate belief permissivism in response to them. Together I think that they help articulate some of Wittgenstein's most interesting reflections on religious language and practice, and explain O'Hara's confusion of faith-based commitments with beliefs amounting to knowledge. While what Sessions and Kitcher each propose seems possible and self-consistent, many people will treat it as a matter of theological correctness that one must reject Session's gap between faith being a matter of conviction *rather than* certainty. They will reject *both* of Kitcher's two self-restricting choices along with it. But what is going on in such cases of discontent with a Left-Bank response seems to be that a more radical fideism is asserting its orthodoxy over a more moderate and self-aware faith venturer. The more radically fideistic one's conception of faith, the more that Session's liberal re-interpretation and Kitcher's self-limitation from knowledge-claims are taken as unnatural impositions of liberal theology, or of philosophy, or of Enlightenment reason: radically fideistic impulses are evident as they declare any such self-restriction from self-ascribing religion-specific 'knowledge', as most unnatural oughts.

Yet in the face of great contrariety in the domain of religious/areligious world views, culturally specific testimonies seem like poor candidates for proper basicality, or for a 'hinge epistemology' might uphold these radicalizing impulses through largely negative apologetics. Consider also that Wittgenstein's position on criterial uses of language doesn't appear to be as permanently fixed as the hinge metaphor suggests. He does not always lean only on the hinge metaphor, but sometimes speaks more diachronically of a river-bed. Here not just the contours, but what is most stable, and what most fluid, may change: 'The river-bed of thoughts may shift' (OC §324). The river-bed *does* move, but more slowly than the waters which flow upon it.[52]

Kitcher accepts the reasonableness of doxastic faith ventures, but where qualifications termed 'skeptical' by their critics are accepted as part of faith, and not the undoing of faith:

> No champion of any religion should be perturbed by the thought that he or she cannot provide marks that distinguish the preferred beliefs about the supernatural from those offered in rival traditions, or worried by the fact that religious traditions evolve in ways that have nothing to do with truth. To be sure, if religion were a form of knowledge, these considerations might be unnerving. That, however, is to mistake the character of religious acceptance. Properly understood it is a matter of faith.[53]

Belief, or at least its self-ascription by the religious insider, is permitted on this broadly permissive ethics of belief, but not the self-ascription of religion-specific 'knowledge', where this is supposed to add something over and above the idea that to believe that *x* is already to believe 'that *x*' is true. Philosophical reflection and argumentation are valued, but not taken to be the *cause* of belief, nor the end-all in its evaluation by different individuals.

Session's and Kitcher's proposals both acknowledge the risk-taking which Tillich draws attention to. They are both clearly in the permissivist spirit which one finds not only in Lessing but in Voltaire and many others of their time in Europe and the Americas.[54] They respect that reasonableness, and interpersonal permission, often follows *possibility,* not the far different norm of high probability, or inductive strength.[55] They both seem to be to elicit what William James famously describes as 'that spirit of inner tolerance without which all our outer tolerance is soulless, and which is empiricism's glory'.[56] It is consistent with acknowledgement of the 'burdens of judgment' with respect to the holistic weighing of evidence in acquisition and maintenance of their 'worldview beliefs', or what John Rawls called in less doxastic language, their *comprehensive conceptions of the good.* But it does not suppose that, having utilized the burdens of judgement to defend one's own reasonableness, one is free to invoke superiority assumptions and exclusivist responses to religious differences that violate these burdens by placing the basis for claiming such superiority in a tradition, or authority figure external to oneself – that is, by what psychologists term *belief disownership.*

There has been a good deal of study of the virtue of intellectual humility and some have argued that a philosophical account of it must go beyond attentiveness of human weaknesses, to 'owning' one's intellectual limitations, especially in domains of controversial views: '[O]wning an intellectual limitation consists in a dispositional profile that includes cognitive, behavioral, motivational, and affective responses to an awareness of one's limitations.'[57] Owning and disowning beliefs and limitations can be studied psychologically, and not, of course, only with religious beliefs. Owning/disowning have a 'multi-track dispositional profiles', which are increasingly being studied in social psychology through specific markers.

Belief ownership and acknowledgement of risk are conceptually entangled. Faith and risk cannot be dissociated, even if negative religious apologetics typically incorporate discrediting mechanisms explaining the wrongness of teachings contrary to those of the home religion. So, one conclusion of our study is that philosophers of religion, whether 'hyphenated' or not, should incorporate psychology, including bias studies. Another conclusion is that theologians, philosophers, religious studies and CSR scholars all ought to study the 'oughts and noughts' of prescriptively fideistic models of faith, and the rhetorical strategies, which Right-Bankers employ.

The 'oughts' and 'noughts' ('ought nots') which logicians, philosophers, decision-concerning sciences, religious authorities, theologians and others issue, as prescriptions for virtuous thought and conduct, are quite varied, indeed. They often cite independent fonts or sources for the criterion of right or efficacious agency, right or wrong judgement, which they assert as normative over us. This is because they draw from different *telos,* or type of value: moral, epistemic, alethic, salvific value and so on. Philosophers and theologians have often tried to reduce these perceived sources and types of value to a single principle stating the biggest of oughts: for one's best judgement all things considered. This at least admits the need for holistic evaluation evidence, but it does not yet recognize how far the contrariety amongst such candidates for a master principle of thought or conduct is due to taking different things as 'evidence', and using concepts, including philosophical concepts, in different ways. When this is the case, the result is often to impose evidentialist standards or a burden of logically sufficient evidential reasons where they don't belong. Or else it is to pit salvific value in God's eyes, or one's fate come Judgement Day, *against* reason in some untoward way.

I propose instead that the thing which these 'oughts' and 'noughts' have in common is that studying them has everything to do with studying *risk*. The most ardent of theists and atheists I think can agree with this, and that knowledge is self-ascribed in the relative absence of objective risk, not in an agent's simple denial or ignorance of it.[58] I hope to have convinced the reader that both risk-embracing and risk-eschewing normative prescriptions can appear unnatural in the extreme, even to the same individual. This is especially so with regard to one's own religious ideas, which may fail a test of authenticity if not logically supported (held to a standard of scientific or everyday discourse), or again, if not held stably, or whole-heartedly enough (prescriptive fideism).

But the worst response to tensions between reason and faith perhaps is to embrace just one of these unnatural prescriptions, while entirely denying all others. This raises a paradox which in turn invites Conflict: on the one prescription (for teleological suspension, or self-privileging in knowledge attributions) one's religious beliefs threaten to lose their reasonableness; on the other prescription (for public evidence and non-circular justification), their specialness! This is a choice which, if taken as forced, blinds us to ground for intellectual humility with respect to self-attributions of religious truth and knowledge, as well as to how my own doxastic risk-taking may result in epistemic injustice to others. To choose only one of these unnatural prescriptions and not to find complementary value in the other leads to loss of the grounds for reasonable disagreement, since reasonability is now cast in terms of one radical ideal theory or the other. For the evidentialist this choice forces them to deny reasonable disagreements even in domains of controversial views. For the theist it threatens loss of *both* the reasonableness *and* of specialness of their doxastic faith venture. In contrast both to the attitudes of an O'Hara and a Russell as Wittgenstein uses them, our study would spur more theologians to engage with the relevance of CSR and bias studies, and to re-evaluate models of faith in relationship to the risk-taking they prescribe.

I have not said that Wittgenstein clearly or unambiguously preferred his first, or Left-Bank response to Lessing's problem, the 'ugly, broad ditch'. But I have tried to show that it fits better than the Right-Bank response with the conceptions of faith scholars have discerned in his writings and that in LRB he is not wrong to be critical of fudges and 'blunders' in Right-Bank responses. And I have tried to give the reader additional reasons to think that *enkratic* or Left-Bank responses contribute far better to inter-faith dialogue, and to comparative philosophy of religion, than can *akratic* or Right-Bank responses. They support doxastic responsibility and intellectual humility, and do not 'disown' this risk of committing hermeneutic or testimonial injustice towards religious outsiders though our attitudes towards them and towards religious differences more generally.[59]

8.5 Conclusion (or, an unapologetic postscript)

We have now seen that Wittgenstein's focal concerns in LRB are not purely descriptive but also reflect his own fideistic orientation, and the value he attaches to religious practice. Philosophy should observe and support our natural tendencies to religious thought and practice, something that Russell is seen to discount, and O'Hara to distort. Russell's fault, one might say, is his *philosophical* correctness: religious beliefs and those who hold them will be evaluated like normal beliefs according to symptomatic or inductive reasoning, without first reflecting upon differences in grammatical usage and the wide range of doxastic and sub-doxastic states that might be characterized as religious. O'Hara's fault is his ascension to this evidentialist demand, and his conflating of subjective certainty with objective certitude in his biblical beliefs. This way of 'cheating himself' includes not only what Wittgenstein seems to regard as the *logical*, but also the *theological* incorrectness of O'Hara's religious epistemics, in so far as O'Hara speaks as if he 'risks little' in his self-attributions of religious knowledge.

O'Hara's apologetic project leads him to unnecessary conflicts between his reason and his faith. Wittgenstein is far closer to Lessing and Kierkegaard than to either Russell or O'Hara. On my reading Wittgenstein is drawn to an *enkratic* fideism, or the first of his two ways. Wittgenstein's response to Lessing's paradox is quite akin to Lessing's own, when the latter confesses: 'This, then, is the ugly broad ditch which I cannot get across, however often and however earnestly I have tried to make the leap.'[60]

But if Wittgenstein means to devalue the enterprises of natural theology and its contrary, disproof atheism, his conception of faith would seem perplexing, and itself an extreme rather than moderate religious fideism. Does not faith seek understanding, and is it enough for O'Hara to *confess* belief without analysing it? While the interpretation of Wittgenstein offered in this chapter explains his distaste for religious apologetics, I would have to take issue with Wittgenstein if he means to completely devalue the work of natural theology. But Wittgenstein of course *himself* reflects philosophically on his beliefs, and like Lessing, Kierkegaard and many others, apparently finds much value in philosophical reflection upon religious language. So, the more moderate thing a Wittgensteinian might say is that one can sometimes benefit from the study of formal arguments, but that their study need not assume that a rationalistic, propositionally focused approach is overall fitting to one's subject matter.[61]

Not all of our natural thoughts are wonders: many deserve to be scrutinized and moderated or let go. And not all seemingly unnatural 'oughts' are blunders: faith *is* quite different than belief acquired inferentially and measurable by degrees of credence. And faith is valued precisely for these reasons.[62] But taking inductive patterns and general facts about human beings seriously, however natural it may seem to be to dismiss them as misleading, provides us a mirror on our own biases. Together with its special concern for Right-Bank theologies and apologetic projects as especially problematic responses to Lessing's problem, this chapter has generally supported the soundness of one of contemporary McCauley's key claims: that while resisted as 'unnatural' impositions on our doxastic practices, inductive norms of evidence 'often provide powerful tools, and serve as vital checks on biases'.[63] While we rarely want to face our cognitive limitations, risk-aware theologians and philosophers acknowledge that 'bias is not a childhood phenomenon that inevitably gets revised and replaced through maturation but a cognitive default that may exist throughout life'.[64]

Notes

1 McCauley 2000, 62.
2 Rush Rhees, a biographer of Wittgenstein, recalls this complaint, and comments thusly on it: '*"Russell and the parsons between them have done infinite harm, infinite harm." Why pair Russell and the parsons in the one condemnation? Because both have encouraged the idea that a philosophical justification for religious beliefs is necessary for those beliefs to be given any credence. . . . Religious beliefs are not analogous to scientific theories, and should not be accepted or rejected using the same evidential criteria*' (1981, 101).
3 Plant 2012, 206, note 58. This is also why what has become known as Wittgensteinian fideism in the literature is weakly prescriptive as well.
4 Wittgenstein 1967, 58.
5 A methodological balance must be found between an emphasis on common form and on distinct ideas and practices, as I argue more explicitly in Axtell 2022a. Such a balance also better allows the relevance to CSR of the wide range of models of faith (or theological methods) which people adhere to.
6 Carroll, this volume, draws on the 'Remarks on Frazer's *Golden Bough*' as a key source for an 'instinctivist' conception of religiosity.
7 McCauley 2000, 65–6. One concern I have is that on the modular view of mind that innatists emphasize, context biases in the uptake of specific theologies and models of faith will remain under-appreciated. Biabanaki (2020) explains, 'Cognitive science of religion researchers divide cognitive biases into two types: content biases and context biases. . . . [W]hile content biases can explain why some religious beliefs are memorable and widespread, they cannot explain why people are committed to and believe in the reflective religious beliefs (theological beliefs) in a particular culture. In contrast, context biases play a central role in the prevalence of theological beliefs. The context biases are shaped by human cultural interactions and act on theological beliefs within a particular culture' (3, 7).
8 2013, 7.
9 779.
10 McCauley and Cohen explain, 'Cognitive theories of religion do not hold that the mind contains a specific module for religion. Instead, most hold that religion exploits a diverse collection of cognitive inclinations in the minds of human beings that enjoy neither a logical nor a psychological unity. The upshot of this analysis is that, cognitively speaking, religions are like Rube Goldberg devices, which is to say that they are exceedingly complicated contraptions calling on all sorts of psychological propensities that are, otherwise, usually unlinked' (2010, 782).
11 Wittgenstein's example of the day of judgement belief, and the passionate appropriation of it which Wittgenstein associates with 'genuine' faith do not fall easily into Boyer's or McCauley's approach, since they find William James to overemphasize emotions, and would likely make a similar criticism of Wittgenstein's account. But while McCauley and Cohen insist that they are not dismissing a role for emotions in religious uptake, other CSR scholars including some more Humean in their approach, research the role of emotions more explicitly (Scott-Kakures 2001; Collier, 2014).
12 238.
13 'Theological' here is inclusive of teachings stemming from narrative testimonies in different or overlapping testimonial faith traditions; 'models of faith' are often

expressed narratively through exemplars (in middle-eastern religions, Abraham, Job, Moses, for example). While Wittgenstein dislikes the term, his example of the Day of Judgement is one of a 'theological' belief. See the religious/theological belief distinction, endnote 4, earlier.

14 Nor is Hume's account of emotion overly narrow, since he distinguishes indirect from direct emotion and does not reduce emotions to feelings. On the contrary, 'Hume makes a number of important contributions to our understanding of the causal antecedents, psychological mechanisms, and behavioral effects of the indirect passions' (Collier 2011, 15).

15 CV 37-38. While this is an exceptionally strong statement of the passions as motivating acceptance of the 'message' of narrative gospel teachings, emotion arguably plays a significant role in *all* of the different conceptions of faith which scholars have distinguished in Wittgenstein's thought.

16 Affectively-driven evaluations thrive on conflation between the *de dicto* and *de re* standing of evidence, that is, roughly, between subjective conviction and objective certitude. This for me is a big part of the problem, and amongst the best ways to connect Wittgensteinian philosophy of religion and contemporary CSR.

17 Wittgenstein 1967, 56.

18 CV, 64. Reflections such as these suggest the distinction between believing *that*, and believing *in*. Louis Pojman (1986b, 467) relatedly argues that faith can best be understood as hope or hopefulness, rather than belief. See also Howard-Synder 2018.

19 McCauley and Graham (2020, 14). So, any cognitive naturalness that scientific reasoning may take on 'is a decidedly practiced naturalness' (286), and this means that together with the counterintuitiveness of its epistemic norms, scientific thinking starts out with a natural disadvantage to maturationally natural ideas, in terms of their adoption and transference.

20 15.

21 16. Experimental psychology indicates a human penchant for relying on their maturationally natural cognitive systems 'even when their deliverances are thoroughly contrary to the norms of deductive and probabilistic inference' (McCauley 2013, 237). Religious metaphysics 'overwhelmingly square with the deliverances of our maturationally natural cognition' (224).

22 The radical counterintuitiveness of certain prescriptions associated with theological correctness as Catholics, Protestants, or others understand it is not interesting only for evidence of widespread non-compliances, or a return to simpler and less demanding religious expressions. The plurality of 'models' of faith, and the impact that adopting a particular conception has in the development of religious identity needs to be examined also.

23 Wittgenstein 1967, 53.

24 Wittgenstein 1967, 54.

25 Wittgenstein 1967, 57; 59. Wittgenstein also writes, 'We don't talk about hypothesis, or about high probability. Nor about knowing. In a religious discourse we use such expressions as: "I believe that so and so will happen," and use them differently to the way in which we use them in science. Although the temptation is great to think we do. Because we do talk of evidence, and do talk of evidence by experience' (56). Wittgenstein uses 'different' thirty-eight times in LRB, to describe the types of reasons or judgements in religious, in contrast to everyday, scientific, or 'normal' (used twelve times) situations and concept use.

26 Wittgenstein 1967, 58.

27 Wittgenstein 1967, 58–9. A closely related passage from *Culture and Value* provides further evidence of Wittgenstein grappling with Lessing's problem: '"Christianity is not based on a historical truth, but presents us with a (historical) narrative & says: now believe! But not believe this report with the belief that is appropriate to a historical report, – but rather: believe, through thick & thin & you can do this only as the outcome of a life. Here you have a message! – don't treat it as you would another historical message! Make a quite different place for it in your life. – There is no paradox about that!"' (CV, 37). The passage shows that Wittgenstein sees differences between narrative and non-narrative testimony (see Chapters 3 and 4 of Axtell 2019).

28 Wittgenstein was quite familiar with Lessing's work on religion. Thanks to Robert Vinten for pointing out the strong evidence of this in Drury (2018, 18–19; 71 note 95).

29 The existential gap 'involves the "problem of religious appropriation". This ditch concerns the "conditions necessary for an individual's successfully apprehending, accepting, and perhaps even understanding the religious message", which may seem "dubious, strange, or fantastic"' (Benton 2006, 35).

30 4. Briefly, the temporal gap stands between historical evidence and biblical miracles: 'The temporal distance between himself and the time of Christ is problematic, since the proof which was enjoyed in the first centuries of Christianity, the "proof of the spirit and of power no longer has any spirit or power, but has sunk to the level of human testimonies of spirit and power". Thus, "reports of fulfilled prophecies are not fulfilled prophecies; that reports of miracles are not miracles", and finally, that such testimonies are "a medium which takes away all their force"' (31–2). The metaphysical gap frustrates further metaphysical inferences, even from evidence of particular miracle events: 'accidental truths of history can never become the proof of necessary truths of reason' (Lessing 2005, 53). Lessing here points out the distinction between claims about 'what happened (which he now grants for the sake of argument) and conclusions of religious truth derived from them' (Benton 2006, 5).

31 Queloz and Cueni (2020) describes Right Wittgensteinians as denying a critical role to philosophy, and encouraging excessive conservativism. A Left Wittgensteinian view is described as allowing for critical discrimination of practices which merit reasonable confidence and practices which do not. Wittgenstein points out views genuine faith as a 'love' or 'what Kierkegaard calls a passion' (CV, 33; 53). The idea of prescribed certitude already invites these logical confusions and very unnatural 'oughts'.

32 On the one hand, Wittgenstein sometimes construes 'standing fast' in naturalistic terms as 'something animal' (OC, §359). On the other hand, Wittgenstein criticizes religious dogma and authoritarianism for being 'irrefutable and beyond the reach of attack' (CV, 28). In his response to the appearance of such paradoxes we see some of the strongest aspects of what Andrejč terms an *existential* account in Wittgenstein.

33 Richard Kraut (2018, np) writes, 'The Greek terms are *akrasia* ("incontinence"; literally: "lack of mastery") and *enkrateia* ("continence"; literally "mastery"). An akratic person goes against reason as a result of some *pathos* ("emotion", "feeling").' For discussion of a paradox about enkratic, or *continence conditions* on rationality, and about the type of irrationality which epistemic akrasia or incontinence presents, see Lasonen-Aarnio (2020).

34 Our distinction between Left- and Right-Bank Wittgensteinians, and between enkratic and akratic responses to Lessing's problem, I have recently found to be described in other terms in Katherine Dormandy's work, and in particular in her distinction between two defenses of *doxastic partiality* which she calls 'anti-epistemological' and 'epistemological partiality.' The former, which she uses

Kierkegaard to illustrate, is heroic, acknowledging the dissonance that attends stubborn counterevidence to one's beliefs, but supposing that 'noetic struggle makes faith excellent'. The latter, epistemological partiality, by contrast 'eschews struggle and fosters noetic ease'; it aims not just to avoid cognitive dissonance but to annihilate it (2021a, 11).

35 The Intellectual Golden Rule (IGR) as discussed in the literature (Hick, 235; see also Knitter) is very much in the spirit of Gotthold Lessing's famous 'parable of the rings' section in his play, *Nathan the Wise*. So mutualist ethics/epistemics and acceptance of (IGR) seem conceptually linked, as does particularist ethics/epistemics with denial of (IGR).

36 McCauley and Graham (2020, 19) write, 'Narratives organize long trains of events that turn on the causal relations between intentional agents' states of mind and actions and their effects on their own and other agents' states of mind and actions. Fashioning stories along such lines, in which events are woven together on the basis of agents' mental lives and conduct, makes them easy to comprehend and easy to remember.'

37 35.

38 Craig and Harris (2008, np). In Axtell (2019), I argue that Craig's approach appeals to being specially placed to receive truth, but that this response exacerbates problems of religious luck.

39 LRB, 4. The sharpness of his criticism of O'Hara's evidentialist apologetics is explained in part by Wittgenstein's conception of faith, a conception which is *prescriptively* fideistic in the sense that it tends towards an extreme he expresses thusly: 'The point is that if there were evidence, this would in fact destroy the whole business' (2–3). 'What seems to me ludicrous about O'Hara is his making it appear to be *reasonable*' (LRB, 4). The general point for us here is that the more the conversants take their religious acceptances as assertions of historical facts where a standard of evidential support is presumed, the more that genuine contradiction between them will be possible, but also the more unreasonable the religious controversy becomes.

40 The final chapter of *Problems of Religious Luck* (2019) proposed a research programme at the intersection of CSR research concerned with authority assumption in testimonial faith traditions, and an inductive risk account of the ethics of belief. O'Hara ascribes high credence and epistemic justification to his belief, but his internalist account is philosophically unsatisfactory because the proposition is not inductively strong, and the evidence he attests is not of the same kind to make strong inferences. Religious narratives place readers where they can feel what it is like to have certain sorts of experience, but they appear not to situate them well for making truth claims, let alone exclusivist ones, on their basis (Fraser 2017).

41 Michele Palmira's (2018) aptly titled Kinship account aims to account for the relationships between not just between full and degreed doxastic attitudes, but also between *believing* and *not having belief*; between *not having a belief* and *disbelief*; between each of these latter two and *middling or low credence*, etc. Palmira's form of epistemic pluralism would seem to articulate Wittgenstein's point here, and to complement or extend the appealing characteristics of Wittgenstein's account.

42 The term 'teleological suspension of the ethical' enters the literature in Kierkegaard's discussion of the paradox of faith exemplified in Abraham's decision in the *Genesis* narrative to follow divine command in the sacrifice of his son, Isaac, in *Fear and Trembling* (1985). I am extending the concept by showing how fideistic and confessional thought operates by an analogous suspension of epistemic and even logical norms.

43 Luhrmann (2020) uses 'faith frame' in place of conception or model of faith. This goes some distance towards accommodating Wittgenstein's crucial distinction between criterial and symptomatic uses of language. It is not as if these frames are strictly incommensurable, or that they are in logical conflict. If either of those were the case, there would be no sense to Wittgenstein giving such long and careful thought to how they relate, as he does in the LRB. Different frames coexist, but problematically.

44 Mentioning cases like beliefs about transubstantiation and virgin birth, Wittgenstein concedes instances in which there could be 'contradiction' between something of a theological nature stated, and that statement's denial. But this is quite unnatural it seems, in that it presupposes a kind of truth-aptness inappropriate to them, at least in the biblical cases. 'And so if Moore said "I know that this is wine and not blood," Catholics would contradict him' (OC §239). Even if we surmise that Wittgenstein would criticize the value of natural theology, he is not just reflecting a more Protestant view, but noting an important distinction. Father O'Hara is an outlier even amongst Catholic priests for his taking religious rationalism to apply to 'proving' religion-specific or scriptural teachings, or epistemic rationality of beliefs of this sort. This is just the sort of 'proof' that Lessing finds wanted but lacking.

45 Grube (2015, 421). Grube (2005, 359–60) rightly points out that 'Lessing provides a way out of the dilemma between either uncritical thinking and reducing/rejecting positive religion,' a way on which the faithful in testimonial faith traditions do not have to leap the ditch, or 'overstep the limits of what is humanly possible to know in order to maintain positive religion.' Grube argues that this is one of the ways (another is sociological or other (including postmodernist) 'reductionisms' *of depriving the concept of truth of its normative function*. For further discussion of the tendencies of fundamentalists to reduce religious contrariety to a simplistic 'bivalent' relationship between (absolute) truth and falsehood, see Grube (2005), and Grube and Van Herck (2018).

46 Compare Bouwsma (1987, especially 54–8).

47 Wittgenstein, *PI* §109.

48 Tillich 1957, 116–17. Compare Andrejc (2015), which develops ways of relating Wittgenstein's grammatical investigation and Tillich's mature understanding of philosophical theology.

49 Sessions 1994, 68.

50 Wittgenstein's scepticism about this expectation is reflected in his claim that 'the chain of reasons has an end' (*PI* §326). In *On Certainty* we often find Wittgenstein discussing the differences between using 'certainty' as referring to a credence function, and something else, perhaps 'certitude' to refer to faith as an abiding and resilient commitments.

51 Kitcher 2012, 263–4.

52 Wittgenstein by my count uses 'hinge' three times, and 'river-bed' twice in *On Certainty*; but the more dynamic metaphor of river waters and bed may be the more apt one for philosophy of religion, such that we should prefer 'river-bed epistemologies' over 'hinge epistemologies' if these are thought to pull in different directions. Although both are active and not static, there remain important differences in the two metaphors, since unlike maturationally natural ideas never to be seriously questioned, the river-bed emphasizes the diachronic flow or progression of ideas. I will have to wait for another occasion to compare my views to those of Duncan Pritchard and Michael Williams (this volume) on Pritchard's proposed 'quasi-fideism.'

53 However, analytic apologetics arguably confuses the defence of reasonable belief with defence of religion-specific knowledge. When this is the strategy, a 'skeptical fideistic' compromise capable of supporting religious inclusivism or pluralism is bypassed in favour of each home religion having an equal right to assert particular 'knowledge', the resulting logical incompatibilities between such claims, be damned.
54 The inductive risk account of the ethics of belief which I develop in *Problems of Religious Luck* and more recent papers, and my use of the Rawlsian notion of 'reasonable pluralism', is largely neutral to debate over the appropriate role for religion to play in the public square, on the one hand, and to debate between Hickean metaphysical unity. Reasonable pluralism and the study of progressive and degenerative research programs (the Lakatosian 'core and belt' model) as the implication of holistic evaluation, does not demand its critics claim.
55 John Bishop (2007) has also explained how recognition of a right to *moderately* fideistic assumptions is the most philosophically sound and practically effective way to constrain *strongly* fideistic faith ventures.
56 It is consistent with Thomas Jefferson's statement that 'It does me no harm whether my neighbor believes in twenty gods or no god. It neither picks my pocket, nor breaks my leg', and with what Richard Rorty called the Jeffersonian Compromise. William James argued that unless harm is involved, we ought to respect others as the rightful choosers of their own risk, and that 'We ought . . . delicately and profoundly to respect one another's mental freedom [and] . . . live and let live, in speculative as well as in practical things', *The Will to Believe*, 234.
57 Whitcomb et al. (2017, 519).
58 For more contemporary work in religious philosophy which acknowledges risk, see Schwenkler (2020) and the other papers in Lambert and Schwenkler 2020.
59 Risk-aware permissivism which I elsewhere develop as *zetetic responsibilism* (2022a, 2022b, 2022c) acknowledges the normalness of risk-taking, including doxastic risk-taking, in identity fusion.
60 Lessing (2005, 55).
61 Strong religious fideism supports what Choo (2018) terms 'unconfirmed superiority debates' over the one wholly true creed or salvific religious identity, and that is an assumption motivating many or most apologetic projects. Dormandy describes this as 'epistemic elitism – asserting a fundamental epistemic difference between the blessed and everyone else, while advocating epistemic norms that further benefit the former by harming the latter' (2021b, 107).
62 Belief-Credence dualists allege that 'beliefs and credences are two different epistemic tools used for different purposes'. See Axtell 2022afor a comparison of Wittgenstein with contemporary Belief-Credence Dualism.
63 McCauley 2013, 2–3.
64 Szocik and van Eyghen 2021, 22. Thanks to the *Hayek Fund* and the *Institute for Humane Studies* at George Mason University, for funding which aided presentation of an earlier version of this paper.

9

Normative cognition in the cognitive science of religion

Mark Addis

9.1 Introduction

The cognitive science of religion aims to gain insight into many religious phenomena by investigating its cognitive foundations and cultural expressions. Such investigation employs a wide range of methodologies with many of these utilizing approaches and results from other fields of cognitive science. Cognitive science in general remains a methodologically contested area with the consequence that aspects of its appropriate application may be contentious. In the broadest terms normative cognition can be characterized as the processing of symbolically mediated normative information. It involves the acquisition, production, distribution and implementation of social concepts and norms through the cognitive ability to use symbols, concepts and norms as normative cognitive tools. Normative cognition encompasses both the logical features of norms in terms of what is required, permissible or prohibited and the psychology of norms in terms of norm-governed behaviour. In doing so it enables the distinction of normative from normal and good from bad.[1] Normative cognition in religion covers a wide variety of religious phenomena including shared religious practices and beliefs.[2] Knowledge of what is right and wrong, and pure and impure is a guide to cognition and an assistance to predicting probable behavioural ranges in oneself and others. For example, normative cognition has a central role in the observation of religious ritual.[3] Accounting for the nature and acquisition of normative cognition requires a proper understanding of general issues about the nature and status of claims in the cognitive science of religion. Philosophical problems about the nature and status of these claims quite often have clear affinities with ones about the nature and status of claims in cognitive science generally. These affinities will not be pursued here but recognizing them is useful when assessing methodological questions in the cognitive science of religion.

How normative cognition should be understood is related to the important issue of the sense in which the cognitive science of religion is science. The development of cognitive science has blurred the distinction between natural and social science. The question of which of these it most closely resembles has significant implications for how issues about methodology should be approached. A central but frequently unrecognized issue in the cognitive science of religion is that of the difference between adopting an appropriately scientific attitude and favouring scientism.[4] Approaches to the cognitive science of religion which emphasize causal explanation and reduce the significance of intentional explanation tend to view it as having significant affinities with natural science. Such approaches tend in varying extents to explain religious phenomena in terms of causal explanations about individual cognitive processes. Such explanations seek to demonstrate their scientific credentials but as will be argued later some tend towards scientism rather than being natural scientific. In contrast perspectives which accord an important role to intentionality and the social dimensions of religion favour aligning it with social science. Here the general approach regarding the cognitive science of religion as aligned with social science will be preferred.

9.2 Wittgensteinian methodological approaches

Questions about methodology in the cognitive science of religion are importantly related to a number of key ideas in Wittgenstein. Specifically, some central ideas from Wittgenstein are consistent with and support empirical evidence in the cognitive science of religion about the nature and acquisition of normative cognition. Before examining this further the issue of whether the fairly frequent criticisms of Wittgenstein as anti-scientific raise serious difficulties for the application of his thought to the cognitive science of religion should be considered. Whether and in what ways Wittgenstein might be deemed to be anti-scientific is a large and complex debate which will not be entered into here. For present purposes it will be claimed that it is reasonable to interpret Wittgenstein as making some philosophically significant claims[5] rather than any philosophical claims being purpose relative to the particular contexts of philosophical therapy they are invoked in.[6] Further to this it will also be claimed that any potential or actual inconsistency between Wittgenstein's own methodological precepts about the nature of philosophy as therapy and how he actually writes about philosophy does not in itself invalidate the value of particular philosophical points he makes. Taking these two interpretative approaches allow the possibility of being Wittgensteinian without being anti-scientific thereby permitting the utilization of his work in the cognitive science of religion.

Wittgenstein's concept of philosophical therapy is methodologically valuable for the cognitive science of religion as it highlights the value of conceptual analysis for revealing implicit commitments to problematic assumptions in what appear to be unproblematic approaches and positions. This analytic approach includes exposing underlying patterns of argumentation (such as Cartesianism) with the objective of identifying root philosophical errors which have various different specific manifestations. Allied to this is the emphasis Wittgenstein places upon recognizing that the most frequently philosophically misunderstood concepts are everyday ones (such as ability, belief and thought) rather than technical philosophical ones (such as transcendental idealism). His concept of philosophical therapy can be usefully applied to theoretical accounts in the cognitive science of religion with respect to whether these accounts are intended to serve a descriptive (that is to describe how things are), explanatory (to account for how things are) or prescriptive (to state how things should be) purpose. Distinctions between these types of theoretical accounts matter as there is evidence in the cognitive science of religion of cases where such accounts purport to be of one of these types but actually are of another. Descriptive accounts of religious phenomena may not be adequate explanations of these and likewise, explanatory accounts of religious phenomena may not be adequate descriptions of these. The interest relativity of explanations has the consequence that an explanation may be inadequate because it is uninformative in relation to the desired information. For example, consider the difference between how a robber and a priest would regard a robber answering a question about why he robs banks by saying that is where the money is.[7] Prescriptive accounts of religious phenomena which prescribe how these phenomena should be explained or interpreted (such as specifying particular kinds of causal explanation) can be problematic. This is because there can be a temptation to explain or interpret empirical data about religious phenomena in ways which accord with particular theoretical accounts rather than seeking to ensure that theoretical accounts provide illuminating insight into this empirical data. A consequence is that such prescriptive theoretical accounts may lack genuine explanatory power or more colloquially put they offer explanations which are not really explanatory (such as the idea that paradigmatic cases of religion involve modestly counterintuitive representations[8]).

The extent to which particular theoretical accounts of religious phenomena are adequate explanations matters because explanatory frameworks are linked to the theoretical positions they are embedded in or stem from. The nature of explanations also matters due to their connection with the aim of employing methodologies in the cognitive science of religion which permit the development of empirically testable theories. Empirically testability is a crucial differentiator between science and scientism.[9] Careful attention to the distinctions between description, explanation and prescription in theoretical accounts enables the identification of metaphysical commitments and armchair psychology involved in the formulation of scientific theories or research programmes. Such identification enables an assessment of the extent to which such theories or research programmes can be reformulated to enable methodologically sound psychological investigation (such that about the processes through which religious cognition is learnt). The methodological use of Wittgenstein in the cognitive science of religion offers the potential to ground theories and research programmes in a way which better integrates and aligns the language, practice and theoretical discussion of religion. In doing so this firmly avoids the not infrequently espoused but problematic position that an insider perspective on religion is necessary to study it properly.[10]

9.3 Methodological approaches in the cognitive science of religion

A general examination of issues about appropriate methodological approaches for the cognitive science of religion is required to understand those specifically pertaining to normative cognition. The sustained and systematic disagreement over appropriate methodologies in the cognitive science of religion has theoretical and empirical significance. There are a number of reasons for this:

- Various disciplines contributing to the cognitive science of religion differ in terms of language, concepts, methodologies and perspectives. Determining degrees of linguistic and cognitive equivalence between disciplines can be complex.
- Under-determination of methodological approaches in the cognitive science of religion by the currently available empirical data on religion. This is not to say that all methodological approaches are underdetermined by the empirical data but this is a sufficiently important general issue.[11]
- Methodological disagreement caused by selecting methodological approaches which are consistent with other desired attributes in cognitive science of religion theories. For example, a desired attribute could be the ability to collect particular kinds of psychological data.
- Methodological disagreement over whether the cognitive science of religion has most affinities with natural or social science (as previously discussed). This disagreement has direct implications for whether the cognitive science of religion should approach the study of religion from the perspective of understanding it entirely (or largely) as a cognitive phenomenon rather than as a cultural phenomenon.
- Methodological disagreement arising from divergent philosophical assumptions or commitments in the philosophies of cognitive science and mind.

Philosophical analysis is essential for identifying suitable methodologies underpinning effective interdisciplinary research programmes in the cognitive science of religion. In the absence of such methodologies these research programmes run the risk of becoming degenerative ones.[12] As considered earlier quite frequently the philosophical assumptions or commitments informing the selection of appropriate methodologies in the cognitive science of religion are not readily apparent so philosophical analysis is required to expose them. For instance, philosophical assumptions may be hidden in substantive claims about particular kinds of data correlation religious phenomena show. Many theories in the cognitive science of religion have methodological commitments which have important but often unrecognized implications for the forms accounts of meaning and symbolization should take. Similarly certain approaches to meaning and symbolization lend themselves to particular kinds of methodological perspectives in ways which are not always readily recognized. In what follows there will be analysis of why the influential methodological approaches of reductionism and privacy are theoretically and empirically unsatisfactory for understanding normative cognition in the cognitive science of religion.

9.4 Reductionism and its problems

Just as it is in cognitive science more broadly reductionism is an important sort of methodological approach in the cognitive science of religion. A general motivation for this approach is the position that the cognitive science of religion has significant affinities with natural science although as has been argued there are good reasons for regarding it as more properly aligned with social science. This position of affinity with natural science supports and is quite often allied with the view that reductionism is taking a properly scientific approach (with the model of physics often being taken as an ideal). Reductionism is supported by and linked to a common emphasis in the cognitive science of religion that all good explanations are causal in nature.[13] In particular there is a strand within the cognitive science of religion which aims to provide a reductionist account of religious phenomena which explains them in purely cognitive and non-cultural terms. These methodological adherences to reductionism and explanation in purely cognitive terms result in a commitment to explanations in terms of methodological individualism.

There are substantial philosophical grounds for disputing the view that reductionism in the cognitive science of religion is taking a properly scientific approach particularly when questions of explanatory power are considered. The general philosophical objection to this kind of reductionism revolves around levels of explanation. Being clear about appropriate levels of description and explanation undermines much of the plausibility of reductionism. Finding appropriate levels of explanation which are not reductive in an unhelpful way is important. An essential part of doing this is to identify which kinds of explanation particular methodologies are aiming to satisfy. For example, the claims about individual cognitive systems each with specific functions termed modules made by Barrett (2004) and Whitehouse (2004) involve problematic reductionism. They claim the mind consists of a hypothesized module with shared interconnected submodules with religion being a consequence of the operation of these submodules rather than being a module in itself. Cognitive architecture is taken to be a collection of mental tools involving submodules.[14] For Barrett and Whitehouse more experimental empirical data is required for the discovery and validation of a successful model of this. Their accounts both presuppose and support a causal model of scientific explanation because religious cognition is taken to be a result of the causal operation of the submodules. However, there are no metrics for what constitutes a module so rather than this being properly scientific it is a case of armchair psychology. The emphasis on modules moves the notion of intentionality and agency from individuals to mechanisms, thereby engaging in reification. It follows that this is reductive explanation which is not particularly explanatory.

Due to the range of reductionist approaches in the cognitive science of religion only selected important methodologically individualist approaches favoured by Boyer (2001), Pyysiäinen (2003) and Sperber (1996) relevant to the understanding of normative cognition will be considered here. As exemplified by Boyer, Pyysiäinen and Sperber a core tenet of the cognitive science of religion is that all religious thought and action overwhelmingly depends upon the utilization of perfectly ordinary forms of cognition available to all normally equipped people. For instance, Boyer argues that there is no need to assume that processing religious thoughts involves a special and unique way of functioning that is not shared by any other cognitive functions.[15] Thoughts and actions are characterized as natural if they possess attributes which rely on non-cultural foundations.[16] What these methodologically individualist approaches demonstrate is a commitment to notions about the privacy of mental representations[17] and as will be argued later there are good Wittgensteinian grounds for regarding this as problematic. In opposition to common presuppositions in both religious studies and anthropology the cognitive science of religion claims that understanding religion does not necessarily require extensive and detailed scrutiny of particular cultures.[18] Patterns of individual and communal religious phenomena should be explained in ways which permit the formulation of testable theories about their non-cultural foundations. For example, the epidemiology of representations ensures that although particular religions can disappear religion does not and new religions routinely appear in human populations.[19] Another unsatisfactory consequence of the favoured methodologically individualist approaches to cognition is the elimination of the social aspects of cognition with the result that the latter have to be theoretically reintroduced later in complicated and not particularly persuasive ways.

An important instance of an attempt to give the cognitive science of religion non-cultural foundations is the claim that norms in humans which have emanated from communication show remarkable similarities across individuals and that the distributions of such communicated norms demonstrate that cultural representations have a propensity to be transmitted if they have certain properties. Boyer and Ramble (2001) argue there is cross-cultural evidence demonstrating there is swift and precise recollection of the modestly counterintuitive representations found in religion. From this evidence a group of cognitive science of religion theories have claimed that paradigmatic cases of religion involve modestly counterintuitive representations. This is used to characterize a concept of religion which is grounded in individual cognitive capacities and largely free from particular cultural dependencies. Proponents of this approach maintain that the concept has explanatory and predictive power. However, a serious limitation of the characterization is that modestly counterintuitive representations are a very weak defining condition for religion since there are many non-religious modestly counterintuitive representations such as famous cartoon characters like Mickey Mouse. This limitation is emphasized by the fact that even some of the representations which Boyer and Ramble characterize as religious due to their modestly counterintuitive nature (such as zombies) are not obviously religious in character. What the weakness of this defining condition indicates is that it is not clear this approach has the capacity to fully explain religion especially with respect to its ritualistic dimensions. Consideration of these particular examples of methodological individualism illustrates the general importance of considering methodological approaches which accord appropriate weight to the place of culture in normative religious cognition.[20] According appropriate weight to the place of culture accords with the approach favoured here of assigning an important role to intentionality and regarding the cognitive science of religion as aligned with social science.

9.5 Reductionism and the commitment to privacy

Reductionist approaches favouring methodological individualism, such as those just considered, involve an implicit commitment to the concept of private mental representations. In passing it should be observed that these approaches have presuppositions and contextual framing which makes it unlikely that their proponents would recognize themselves as having a commitment to private mental representations. Generally the commitment which reductionist approaches favouring methodological individualism have to private mental representations is a consequence of how these approaches account for mental representations acquiring and exercising normativity. For example, Boyer[21] claims that thoughts are invisible and unobservable with the consequence they must be inferred. This clearly commits him to the view that the mental representations of others are not directly accessible so inferences about these are constantly required. Having taken this position it is then natural to develop it by attempting to systematically explain the kinds of inferences involved in religious phenomena. However, there is no empirical evidence from psychology that individuals constantly make inferences about the mental representations of others as opposed to doing this as and when required by the situation. It follows that if this position is at all plausible it must be the case that such inferences about the mental representations of others are being made unconsciously. The problem with taking the approach that such inferences are unconscious is that it is no longer clear how Boyer's position about inferences can be empirically tested in a methodologically sound psychological way and this suggests he is tending towards scientism rather than natural science.

There is a vast and complex literature on interpretations of Wittgenstein's private language argument. However, these readings share the view that there are problems with the intelligibility of the concept of private mental representations as this requires a methodological commitment to the intelligibility of the concept of a private language and that no such language is possible. For present purposes it will be assumed that an interpretation of Wittgenstein which offers plausible reasons for denying the existence of a private language can be constructed without going into detail about precisely what form it should take. Boyer's position and ones which are committed to the same methodological precepts about the mental representations of others not being directly accessible can be objected to on grounds raised by Wittgenstein's private language argument. Wittgenstein argued that it is mistaken to claim that individuals can directly observe mental representations through introspection while others can only make inferences about mental representations through bodily and behavioural manifestations.[22] It follows that it is also erroneous to maintain that since knowledge of the mental representations of others cannot be acquired by introspection it can never be as certain as knowledge about one's self. Another crucial problem with Boyer's commitment to the privacy of mental representations is that the intersubjectivity required for normative cognition has to be theoretically reintroduced instead of it being present from the start. This is because although norms are mental representations they are quintessentially shared and thus cannot be private.[23] Assuming Wittgenstein's criticism of the concept of private language and related notions to be correct there are significant consequences for the kind of philosophical approach to a theory of meaning which should be sought.

9.6 Normative cognition

The preceding discussion of problems associated with reductionist methodologies in the cognitive science of religion for understanding normative religious cognition suggests alternative methodological approaches which give due weight to the importance of culture. These latter approaches recognize that religion is closely attached to the whole social world it is embedded in with the consequence that the richness and complexity of religious phenomena should be suitably articulated with reference to this. Following on from this understanding the acquisition and exercise of normative religious cognition requires an appreciation of both its cognitive and cultural dimensions and an acknowledgement that it lies somewhere between these two dimensions.[24] Accounts of and research programmes about the acquisition and nature of normative cognition in religion are related to and dependent upon general accounts of normative cognition. It is thus important to identify features of an appropriate explanatory framework for normative cognition which can be appropriately refined for normative cognition in religion. Following on from earlier discussion an appropriate explanatory framework cannot treat normative cognition in a methodologically individualist way and must take the social dimensions of meaning and symbolization into account. This explanatory framework must also consider that normative cognition in general and in religion in particular involves implicit learning.[25] This importance of implicit learning in these processes is supported by the fact that evidence from psychology shows that most learning is implicit. Most cognitive constructive processes are unconscious as they are neurologically embedded and operate automatically.[26]

One way in which implicit learning can occur is through the process of mimesis and this raises the question of whether an account of the acquisition of normative cognition in religion solely based upon mimesis is sufficient. One suggestion for the neurological processes involved in mimesis is that these utilize mirror neurons. It has been argued that mirror neurons have the potential to provide some empirical explanation of how mimesis might function.[27] In contrast Keestra (2008) claims that mirror neurons are used to explain too much. This is because neurons in themselves are transmission mechanisms which cannot imitate anything and there are potentially many other imitation mechanisms such as high-level inferences from observation. Another problem is the adequacy of explanations in psychology and neuroscience which invoke the imitative and empathetic abilities of mirror neurons but without properly accounting for the attributes of mirror neurons which endow them with the ability to imitate or empathize.[28] These criticisms of explanations involving mirror neurons are well justified and so much more theorization of the idea of mimesis is required for a proper explanation of it. Aside from the particular issue of mirror neurons the general point is that mimetic mechanisms in themselves are not an adequate explanation of the acquisition of normative cognition in religion and other elements are required for a fully satisfactory explanation.

In Donald's account of mimesis there is a cognitive craving for stability and familiarity with experience contributing to this acquisition of this.[29] A loss of familiarity and stability is disorientating and often leads to a temporary experience of meaninglessness. The emphasis Donald places on stability and familiarity illustrates an important general point about the cultural dependence of normative religious cognition. Normative cognition in religion cannot result solely from mimesis because it requires some reasonably well-defined form of cognitive content to operate upon. An adequate account of normative religious cognition needs to provide a cognitive and epistemological account of how cultural dimensions of normative religious cognition become internalized in individuals. Some proposals for the cognitive processes involved will be discussed here and later on the epistemological aspect will be covered with respect to Wittgenstein's conception of world pictures. Cultures create complex shared worldviews that balance the needs of large populations. Most religions utilize art, ritual, public spectacle and specific cultural practices to sustain a world view and are aimed at cognitive governance. Cultural worlds are collectively and unconsciously created with a variety of mechanisms being suggested for this creation.[30] Despite this unconscious creation there can be conflicts between different kinds of normative cognition.

Donald argues that a world view offers a self-complete cognitive framework for living. He maintains that religion is the same kind of system as other systems which transmit worldviews. Donald claims that it is not really possible to have a meaningless world view and that meaning is always framed by a world view.[31] For Donald cognitive engineering is the deliberate expression of a world view and can be contrasted with the sense in which a Wittgensteinian world picture usually goes unnoticed because it is the inherited background. On his account worldviews evolve through cultural stages which have a cognitive dimension in a process of cultural learning. The mythic stage is a traditional form of human culture in language-based cognitive governance by standard version narratives which Donald characterizes as projective meaning in the mythic. The theoretic stage is based on the gradual assertion of the dominance of analytic thinking leading to de-mythologization which Donald identifies as reflexive meaning in the theoretic stage.[32] As will be seen later Wittgenstein's concept of a cultural form of life provides an epistemological framework for the cognitive processes in Donald's conception of the mythic and theoretic stages. Having identified the importance of cultural learning in the acquisition and exercise of normative cognition it is now appropriate to turn to a discussion of a conception of social meaning which could support this kind of cognition.

9.7 Social meaning and shared belief systems

Wittgenstein's later philosophy of language can be used as a basis for developing an account of meaning that persuasively explains its collective nature. Such an account is required to satisfactorily explain the collective aspects of normative religious cognition (such as rituals) including their propagation and transfer. What is crucial here is providing a satisfactory explanatory framework for what goes on between individual minds in the acquisition of collective knowledge or, alternatively put, to adequately account for the acquisition of religious intersubjectivity.[33] Wittgenstein's social approach to meaning fits well with empirical data about the social aspects of meaning (such as that about the processes of cultural transmission), normative cognition and social facts. This is consistent with the increasing emphasis in cognitive science of religion research on accounting for religion in sociocultural terms with due weight being accorded to social processes.[34]

Social facts are linguistically expressed objects of shared intersubjective knowledge. Social meaning and reductionist approaches to accounting for social facts offer conflicting epistemological and ontological accounts. A full resolution of these disputes will not be attempted here. However, there will be sufficient clarification of these issues to securely ground an account of normative cognition. Searle (1995) and Sperber (1996) will be examined as their approaches exemplify the divergences between the social meaning and reductionist approaches to accounting for social facts. Searle's approach originates from Wittgenstein's ideas about social meaning. For Searle social facts are any facts involving collective intentionality which are irreducible to psychological facts, structures and processes although they are both dependent upon and influence them.[35] The objectivity of social facts is constituted by their independence from any individual, thought or will. Institutional facts clearly illustrate this as they are social facts of the general form 'X counts as Y in context C' such as a twenty-dollar bill counting as a monetary token in the United States.[36] In contrast Sperber (1996) disputes whether social facts exist because of their collective acceptance offering an account of them via causal chains with alternating psychological and environmental links. He argues for a cognitive science account about the transmission of culture which claims that culture results from the causal interaction of mental and public representations. The majority of cultural change stems from alterations in the distributions of communicated beliefs in societies but these beliefs are only part of the whole set of cultural representations.[37] Sperber's account is subject to the problems about methodologically individualist reductive accounts and of adequately accounting for intersubjectivity due the commitment to the privacy of mental representations discussed earlier.

Normative cognition linked to social facts in Searle's sense as being any facts involving collective intentionality. It is important to develop a suitable philosophical framework for connecting normative cognition and social facts with part of this being an account of socially shared belief systems. Social facts can be included in Wittgenstein's account of meaning being socially constructed by a community of language users. Such inclusion enables his ideas, especially those from *On Certainty* about world pictures, to be usefully employed to clarify the ontological and epistemological status of social facts in a way which does not involve reification along with articulating their significance for normative cognition. Discussion of social facts naturally leads into the wider concept of socially shared belief systems which stems from it. Socially shared belief systems concerned with the ideological bases for the distribution and exercise of power are large systems of social facts whose functions involve the generation of norms.[38] Importantly religion is a socially shared belief system whose intentionality is clearly evident.[39] As will be seen later Wittgenstein's world pictures serve the same explanatory functions as Sinha's socially shared belief systems although there are some differences in what underpins this explanatory function. An adequate account of the epistemological and ontological character of socially shared belief systems must avoid reification of these systems or their aspects.

Such an account can be usefully developed by considering the types of explanatory functions in socially shared belief systems:[40]

- Existential – explaining how the living and the dead form a single community
- Ideological – explaining why is power distributed and exercised in the way that it is
- Metaphysical – explaining how is the universe ordered and how a community has agency in it

These types of explanatory functions provide a framework for identifying and articulating what purposes particular aspects of socially shared belief systems serve and why they are important to the communities in question.

9.8 Forms of life

It is well known that Wittgenstein's account of social meaning is grounded in his concept of forms of life with the latter having a role in explaining the relationship between normative cognition and socially shared belief systems. As with much of Wittgenstein his concept of a form of life is highly contested with an extensive literature. Although there are very few usages of the expression 'form(s) of life' in his writings it is plausible to think that Wittgenstein employed the phrase in two distinct ways.[41] One use is to summarize the biological aspects of human nature in the sense of the common human way of acting, namely, that which is particularly and universally human. The cognitive science of religion's description of natural cognitive capacities shared by all normal people that underpin religion is consistent with Wittgenstein's idea about certain things being natural given a particular biological form of life. The other use is to refer to the cultural aspects of human nature in the sense of stressing the differences between societies. These aspects are broadly concerned with practices in ways which encompass both anthropology and sociology. The cultural forms of particular religions are consistent with Wittgenstein's idea about certain things being natural given a particular cultural form of life. Both strands of forms of life rest upon the very general facts of nature which are the background stabilities of the natural world. For Wittgenstein these very general facts of nature impose limitations upon which concepts are natural or unnatural to nearly all humans.[42] Wittgenstein's treatment of the notion of forms of life is also consistent with the position prevalent in much of the cognitive science of religion that language is a bio-cultural phenomenon resulting from evolutionary adaptation.[43] Although Wittgenstein never discusses evolution as he would have regarded it as the province of science or psychology rather than philosophy his treatment of biological and cultural forms of life and their interrelationships is clearly consistent with a conception of language as a bio-cultural phenomenon.

9.9 World pictures

The cognitive mechanisms and their associated cultural underpinnings involved in the acquisition and exercise of normative cognition in religion have been considered previously. Here this account of the cognitive mechanisms and their cultural underpinnings will be theoretically strengthened by offering an epistemological framework for this account derived from Wittgenstein's concept of world pictures. Before discussing this, it is important to observe that on his conception of philosophy the cognitive capacities involved in the acquisition and exercise of normative cognition are the province of cognitive science and psychology and not matters for philosophical consideration. The interpretation of *On Certainty* which will be preferred here, although it has been controversial, is that Wittgenstein claims that knowledge is comprised of two broad classes of core and empirical propositions.

Wittgenstein's concept of a world picture is a system of propositions which support one another. A world picture provides the context for meaningful knowledge-claims and for inquiry. It is the background against which other knowledge is acquired and the context in which claims to knowledge are meaningful. World pictures are not based on reason but they are not unreasonable. The craving for familiarity and stability Donald (2004) stresses are crucial factors which relate to Wittgenstein's claim that people are strongly struck by a world picture which is not theirs and why existing world pictures are often unnoticed. In world pictures core propositions are surrounded by empirical propositions which are the result of investigation which is expressed by Wittgenstein's metaphor of the bed and waters of a river.[44] Wittgenstein emphasized that although they take the form of empirical propositions core propositions differ in kind from empirical ones as they do not function as empirically testable propositions. A core proposition may take on the role of an empirical one and vice versa with almost every proposition being potentially revisable. A system of propositions is learnt gradually – some of which are certain and indubitable, while others may be doubted to a greater or lesser degree. Propositions which are certain do not have this status because they are intrinsically obvious or especially convincing. It is rather that they are certain by virtue of those that shift around them.[45] The propositions adopted and whether a particular proposition, such that expressed by a social fact, is plausible depends upon the world picture. A difference in world picture propositions leads to a difference in what is counted as evidence.

Wittgenstein's concepts of biological and cultural forms of life have a role in the creation of world pictures. Certain things are natural given particular biological and cultural forms of life. This helps to account for the noticeable similarities and differences between many world pictures including their constitutive social facts as well as the fact that humans have many shared cognitive norms. Different world pictures are closer and less close to the truth in various respects. Wittgenstein argued that certain world picture propositions could not be revised or rejected because doing so would dismantle the world picture. Sufficiently different world pictures can be incommensurable. As with incommensurability generally arguments from the premises of the world pictures themselves will not be sufficient to resolve disagreement. However, it should be observed that in itself Wittgenstein's account does not provide a complete epistemological framework for the acquisition of normative cognition in religion. In *On Certainty* Wittgenstein offers very little insight into the important issue of how disagreement between incommensurable world pictures can be resolved as his examples only consider persuasion with limited reflection about how persuasion works in cases of this kind. For example, religious conflict can result from incommensurable religious world pictures being brought together through globalization. Major ways of resolving this sort of disagreement are through intercultural dialogue and persuasion. An adequate account of how persuasion works in these instances needs to considerably expand upon the small basis that he provides by recourse to other ideas. In particular although Wittgenstein uses some religious instances,[46] he only addresses the significant question of the relationship between altered normative cognition (such as through religious conversion) and changed world pictures (such as becoming religious) in a very restricted way. Here empirical cognitive science of religion data from anthropology and sociology has an important role to play.

The cognitive science of religion idea holds that the acquisition of normative religious cognitive has far greater affinities with acquisition of a natural language than the gaining of the abilities and knowledge to necessary to undertake scientific work. What underpins this idea is the view that acquiring religious knowledge and norms frequently occurs without explicit teaching as people are born into religious and linguistic communities and is thus an instance of implicit learning.[47] The notion of acquiring knowledge from the community which one is born into is arguably usefully expressed by the conception of a world picture. Wittgenstein claimed a world picture 'is the inherited background against which I distinguish between true and false'.[48] World picture propositions are not explicitly learnt but it is possible to subsequently find out which propositions these are.[49] He argued that the accumulated knowledge of a culture is a collective human achievement.[50] Wittgenstein claimed that learning was based on accepting the authority of a community and that adults as well as children must take much knowledge on trust.[51] Accepting much knowledge on trust allows humans to undertake investigations which modify some of their beliefs.[52] Common religion is one traditional way of defining a community of deep trust. The concepts of a world view[53] and of a shared social belief system[54] serve an equivalent function as shared cognitive worlds. This epistemological framework of world pictures provides a secure grounding for accounts of the cognitive mechanisms involved in the creation, acquisition and transformation of shared cultural worlds which underpin the acquisition and exercise of normative religious cognition.

It has been argued that a philosophical account of the role of normative cognition in the cognitive science of religion is important for furthering theoretical and empirical research in this area with Wittgenstein's work having a demonstrably valuable contribution to make to this. Although a number of unresolved problems about the precise nature of the cognitive and social processes involved in normative religious cognition remain the groundwork for future research agendas can be laid.

Notes

1 Kelly and Davis 2018, 58.
2 Jensen 2010 and Jensen 2013.
3 Boyer and Liénard 2020.
4 De Ridder 2018.
5 Such as in Hacker 2005.
6 Such as in the strong therapeutic readings of Crary and Read 2000 and Baker 2004.

7 Putnam 1978, 42.
8 Boyer and Ramble 2001.
9 Trubody 2019.
10 Day 2004 and Jensen 2011.
11 Laidlaw 2007.
12 Lakatos 1970.
13 McCauley 1996 and McCauley 2017.
14 Barrett 2004, 4–5.
15 Boyer 2001, 311.
16 Boyer 1994.
17 Jensen 2009.
18 Boyer 2001, Pyysiäinen 2003 and Sperber 1996.
19 Sperber 1996, 9–31.
20 Laidlaw 2007.
21 2001, 98.
22 Wittgenstein 1967b, §246.
23 Sinha 2009.
24 Pachis and Panagiotidou 2017 and Shore 1996.
25 Birch and Heyes 2021.
26 Reber 1993.
27 See, for example, Schilhab 2007.
28 Hacker 2018, 381–5.
29 Donald 2004, 51.
30 For example, see the accounts of cognitive schemas in Lakoff and Johnson (2003) or conceptual blending in Fauconnier and Turner (2002).
31 2004, 49–52.
32 2004, 52–6.
33 Day 2004 and Jensen 2013.
34 McCauley 2017 and Panayotis and Panagiotidou 2017.
35 1995, 26.
36 1995, 31–57.
37 Sperber 1996, 25.
38 Sinha 2009.
39 Donald 2004 and Jensen 2013.
40 Sinha 2009.
41 Addis 2013.
42 1967b Part II, XII, p. 230 and 1980b §708.
43 Heyes 2018.
44 1969 §§96–9.
45 1969 §144.
46 Such as 1969, §§106–7 and 612.
47 1969, §107.
48 1969, §94.
49 1969, §§279 and 152.
50 1969, §§288 and 298.
51 1969, §§159f., 170, and 508f.
52 1969, §161.
53 Donald 2004.
54 Sinha 2009.

10

Brains as the source of being

Mind/brain focus and the Western model of mind in dominant cognitive science discourse

Rita Anne McNamara

10.1 Introduction

The ability to infer another being's thoughts, goals and desires from their observable behaviour is a key skill that humans rely on. Western societies often take this focus on the mind as a given, even to the point that some claim others can be read like a book.[1] Though this is an apparently obvious means of navigating the social world in the West, people from other cultures place more emphasis on the situation.[2] Ethnographies and developmental studies from small-scale societies show more collectivistic communities develop abilities to interpret and predict behaviours based upon observed actions in similar phases across cultures, but the ages at which these abilities appear vary.[3]

Research traditions rooted in Western culture – especially psychology, cognitive science and neuroscience – perpetuate the view that mind perception and operations of the brain are the keys to all social behaviour.[4] However, as these researchers are themselves part of a cultural setting, these theories often fail to take into account how this cultural world view influences the theories, methods and conclusions they develop. The idea that the mind is the locus of all action is in fact a cultural model of mind: a perception/conception of mind that is a product of the culture it emerges from. This chapter provides a brief survey of how cultural models of minds, both formal and informal, influence religious belief in different cultural settings. The chapter starts by describing the Western model of mind in its informal/folk and formal scientific versions, as well as how it emerges in scientific studies of religion with cognitive science of religion and the evolution/cultural evolution of religion. Following this review of the Western model of mind, the chapter next describes alternative models of mind within non-Western traditions. Indian philosophical traditions within Advaita Vedanta, Buddhism and Sāṃkhya present a set of comprehensive, formal theories of mind that remain distinct from Western formal (scientific) and informal (folk) theories. Following this exploration of alternative formal theories of mind, this chapter briefly describes alternative folk/informal theories of mind: perception of agency and plant intelligence in the Americas and Opacity of Mind in the Pacific. Each of these folk theories of minds is followed by a brief description of how they impact belief.

10.2 The Western model of mind: I think, therefore I am

10.2.1 The informal/folk model

A society's folk model of mind forms the basis for naïve psychology, the implicit assumptions individuals make about how other social beings operate in the world.[5] In the Western folk model of mind, individuals operate more or less autonomously, driven by phenomena like preferences, instincts, goals – all of which originate in the mind. Therefore, if one wants to know another being, one must know their mind. The mind is the seat of the self, which is discrete and bound from others and the surrounding environment.[6] Reaching back at least as far as Descartes, the Western model of mind also incorporates an intuitive mind/body dualism with the mind and the body being treated as discrete entities. The mind, and by extension the self, is believed to be located within the physical medium of the brain.[7]

This bounded, discrete mind gives rise to the problem of solipsism – if one's mind is fundamentally separate from all others, how is anyone to know that other minds do in fact exist? The solution to this solipsistic problem is the belief that thoughts and other internal mental states are the mediators between the mind and the physical world. Thus, if one can observe another's behaviour, because these behaviours are caused by the mind, they can be taken as directly indicative of that mind's contents. Focus on interpreting behaviour through the lens of intent further recapitulates this focus on the importance of individual mental states as the driver of action.[8] This leads to an awareness that individuals' subjective perspectives may lead to very different conclusions about the same underlying 'objective' reality. A person's responses to events depend on their thoughts about that event – and therefore their observable actions are inferred to be more indicative of their underlying dispositional character than the situation. This leads to the common Western assumption that internal thoughts and characteristics are more causally important in determining behaviours than situations are, giving rise to reasoning biases like the Fundamental Attribution Error.[9]

This focus on minds within the Western model of minds is ecologically linked to increases in physical, material stability and predictability of the environment (i.e. rule of law, court systems, basic infrastructure etc.) that occurred due to the rise of governmental institutions that foster greater focus on individual rights, freedoms and greater deviation away from strict group norms.[10] As individuals are given more freedom to act based upon individual preferences and values, the mind becomes a rational and cognitively efficient target for understanding another's behaviour.[11] Historically, the rise of Christianity, and Protestantism in particular, appears to have had a large role to play in building this focus on minds by removing the connections to kin and environment that typify the more 'porous' or unbounded minds that interact with environments more fluidly in many non-Western traditions.[12] Sparked by the Protestant Reformation, people began to look for evidence of salvation not in actions (like asceticism and good works), but in the direct, unseen, inner connection to God via belief.[13] This Western emphasis on individualism makes the mind the locus of the self, the self the all-important arbiter of behaviour, and thus behaviour indicative of internal mental states.[14]

10.2.2 The formal model: Scientific study of the mind in cognitive and neuroscience

While folk models of mind are intuited and built up through socialization experience, a formal model of mind constitutes a much more explicit, effortful attempt to explain how minds and beings operate. Within Western society, formalized theories predominantly fall within the domain of science, specifically within psychology, cognitive science and neuroscience research disciplines. Neuro-cognitive scientific approaches place the mind and the associated conscious experience it entails within the physical material of the brain.[15] With the rise of computing, subjective experience has increasingly been understood via metaphors with computing,[16] though more recent models now focus on a predictive brain that combines bottom–up sensory experience with top–down memory representations.[17]

As these scientific approaches to the mind are themselves a product of the Western cultures that created them, they too show the marks of the informal Western model of mind. For example, the ability to infer the contents of another mind is often theorized to be an essential requirement for survival as a social species.[18] As a consequence of the Cartesian dualism within the Western model of mind, these theories typically distinguish between rational thought and other kinds of more physical, experiential processes. Thus scientific study of the mind is discussed within two broad, orthogonal dimensions: (1) the domains of cold/cognitive/agentic abilities versus hot/affective/emotional or experiential abilities and (2) understanding the minds of others versus understanding one's own mind.

Mentalizing, or the ability to infer the contents of another mind and the tendency to use them to interpret one's environment, forms one subset of these abilities. Mentalizing abilities are typically discussed in one of two ways: either the more rational, cognitive 'cold' abilities under theory of mind or the 'hot', affective, emotion-focused empathy. First coined by Premack and Woodruff (1978) as the ability to '[impute] mental states',[19] theory of mind processing typically refers to inferences about others' knowledge states, goals and especially true or false beliefs that may differ from one's own.[20] Empathy, on the other hand, was first introduced into English as a 'process of humanizing objects, of reading or feeling ourselves into them'.[21] Empathy typically refers to the emotional processing components that we use to orient each other to shared emotional and affective experience.[22] This emphasis on shared feeling leads empathy to be associated with altruism, as the literal sense of feeling another's pain drives motivation to alleviate it.[23] Much of the enquiry into theory of mind and empathy across these research disciplines has been through the lenses of dysfunction or development.[24] Theory of mind deficits are the suggested root of autism spectrum neuroatypicalities,[25] lack of concern for others' emotions are considered an important antecedent of psychopathy,[26] and lack of awareness of one's own emotions is a characteristic challenge within alexithymia.[27]

Another dominant perspective within these formal scientific theories of mind, dubbed mind perception, has similarly been divided into two orthogonal domains of agency and experience.[28] Agentic beings are those which have more impact in the world and are less impacted by it; beings like robots that have no emotional lives at all or vast corporations like Google. As they are more in control than controlled, they also hold moral responsibility for their actions. Experiential beings, on the other hand – beings like babies and dogs – are more affected by the world and therefore have less responsibility and need more protection. Some beings, like adult humans, are high on both dimensions; others like a dead person or a person in a vegetative coma are low on both.

10.3 Western model of mind in scientific approaches to religion

The entry point for how Western science goes about understanding religion takes a similarly mind-focused path. God is seen as a hyper-agentic being, and a driving question behind this research becomes: Where does the human ability to conceive of and perceive the divine come from?

Cognitive science of religion places human cognitive systems at the root of supernatural beliefs,[29] with evolutionary and cultural evolution of religion theorists further suggesting this human-perception cognitive system either co-evolved with or was co-opted/exapted to support cooperation.[30] Our ability to infer the minds of others is theorized to be extended to infer the presence and contents of unseen supernatural agents' minds (which are themselves deemed to be non-real entities within this scientific approach[31]). Some theories suggest religious experience is the result of excessive perception of other minds (i.e. hyperactive agency detection)[32] or promiscuous anthropomorphism.[33]

If this hypothesis that religious belief evolved out of our ability to perceive the minds of other (human) beings is true, then it implies that the most effective beliefs about god(s) will be beliefs that make god(s)' minds more like human minds. Neuro-atypical populations who show lower theory of mind are also associated with lower levels of individual belief in supernatural agency,[34] while differences in mentalizing can distinguish religious vs. spiritual believers.[35] The mind of God has also been sought out in brain imaging; these studies show believers activate parts of their brains associated with thinking about themselves when thinking about God, implying we fill in the mind of God by imputing our own.[36]

Though some forms of supernatural agents appear to arise from excessive mentalizing,[37] the vast majority of religious experiences are dependent upon the cultural inputs that give shape to the specific expectations people have about supernatural minds.[38] While mentalizing is related to supernatural belief, this is likely more important in Protestant Christian traditions than in other religious belief systems (because belief is often less important than practice in other traditions[39]).

10.4 Mind across cultures: Ethno-psychologies and social selves

Non-Western models of mind predominantly differ from formal and informal Western notions of mind in their emphasis on social context and relational ties. Relationships as the space in and between individuals rise as the point of focus rather than the individual.[40] Social hierarchy,[41] and the complementary norms that structure behavioural expectations,[42] may limit the extent to which actors' true underlying mental states lead to actual behaviour. For example, cultures that value hierarchy and social order also tend to normatively express a more limited range in emotional displays like anger, contempt and disgust.[43] The following sections highlight several key differences in non-Western formal and informal models of minds to indicate how the mind changes when individual human experience, rooted in the brain, is not taken as the primary frame of reference.

10.5 Other formal theories of mind: Examples from Indian philosophy

Modern, Western, scientific study of the mind is not the only formalized theory of mind; Indic traditions have a history of exploration deep into the nature of mind, body and consciousness that extends back thousands of years.[44] As with most traditions outside the West, these philosophies cannot be readily separated from the spiritual and religious contexts they are enmeshed within.[45] Though these Indic theories of mind stem from spiritual/religious practices, and though their methods are highly introspective and subjective (a stark contrast to the externalized, objective focus in Western science[46]), the theories about mind and consciousness they produced remain systematic and highly detailed. Indic spiritual, religious and philosophical traditions are diverse and complex; few if any sweeping generalizations adequately summarize the entire subcontinent. For the purposes of this chapter, two trends are discussed through three exemplar philosophical traditions: belief in karma and reincarnation as they are implicated in formalized theories of mind developed within Advaita Vedanta, Buddhism (with a focus on the Pāli canon) and Sāṃkhya (the philosophy associated with classical yoga).

Much of the existing work in cognitive science of religion, and to a lesser extent evolutionary/cultural evolutionary study of religion, has focused on Abrahamic traditions, reflecting a pervasive trend that over-emphasizes Western experience in the cognitive sciences and psychology broadly.[47] The current literature therefore may over-emphasize the importance of perceived agency in supernatural causal attributions by overlooking non-agentic supernatural phenomena. Attribution to non-agentic and non-physical causation is a hallmark of Asian karmic traditions, many of which originate in India. Karma is typically formally discussed as an impersonal force or principle of the universe that manifests as consequences of actions.[48] In essence, what you sow is what you reap. Unlike supernatural agents, karma has no mind and cannot be bargained with. In line with the socio-centric self common in collectivistic societies,[49] individually acquired karma can impact others in one's kin group across generations. The only available response is to live through whatever karma has caused to manifest in the present and (especially in the case of liberation practices offered in Buddhism and Sāṃkhya) act appropriately, here and now, to evade accumulating further karmic consequences.

Transmigration of the soul via the constant cycle of reincarnation (*saṃsāra*) is another important belief that goes hand-in-hand with karma.[50] Part of how karma expresses the consequences of action comes through one's *jati*, or social class and conditions of one's birth. One can have a good birth and live a life of good fortune and ease, perhaps with such great good karma as to be born a celestial (nonphysical, divine being with more power and less discomfort than humans) or with such terrible bad karma as to be born a lowly worm. A key component of how this system functions is *duhkha*, or 'unsatisfactoriness', versus *sukha*, or 'happiness/sweetness'.[51] Karmic traditions that embrace the cycle of birth seek to maximize good karma to enable the best possible birth for one's entire family, while practices that target liberation (*mokṣa* in Advaita Vedanta, *nibbana/nirvana* in Buddhism, *kaivalya* 'aloneness/all-one-ness' in Sāṃkhya[52]) seek to exit the cycle by elevating themselves beyond karma and thus beyond re-embodiment.

Turning to our three exemplar Indic philosophy traditions: how does the mind become conceptualized within Advaita Vedanta, Buddhism and Sāṃkhya?

10.5.1 Advaita Vedanta: Material unreality and all-encompassing self

Advaita (non-dual) Vedanta (that which emerges from the Vedas) posits that the only real reality is *Brahman*: pure, contentless consciousness that is the source of all things.[53] In this philosophy, all apparently physical things are in fact epiphenomena; illusions that ebb and flow with change constituted by the same underlying *Brahman* substance. Being a non-dual philosophy, Advaita Vedanta immediately sets a contrast to Cartesian dualism. Its stance of idealism in the claim that *Brahman* consciousness is the actual substrate of reality similarly sets a contrast to modern scientific materialistic realism. Instead, subjectivity, and the ultimate Self (*Brahman*), is the ultimate truth and reality of the universe.

The mind within Advaita Vedanta, though ultimately rooted in pure *Brahman*, has apparently physical, material components. Individual human experience is enabled by the *Ātman* (Self, or true self), which is an emanation or aspect of the greater whole of *Brahman*. This sets the stage for the apparent but false impression of dualism, the sense that the self is distinct from the *Brahman* whole. This illusion of separation and of physical 'reality' in general is created by *Brahman* in a state called *maya* (illusion). Part of what makes *Brahman* and its *Ātman* aspects real is their eternal and unchanging nature; *maya* is not real because it changes. *Maya* is created by *Brahman* in a manner analogous to an illusion created by a magician. Its function is to enable *Ātman* emanations to experience the full extent of *Brahman* through the illusion of change and contrast.[54]

The *Ātman* (and by extension, the *Brahman*) interacts with physical (un)reality via the vehicle of the physical mind. The mind is subdivided into hierarchical components from greater to lesser subtlety, which may be better thought of as a mind-consciousness-perception complex.[55] The most subtle yet still material part of this complex is the *buddhi* (intellect), which is associated with the *antaḥkaraṇa* ('inner vehicle' for *cit* consciousness of the *Ātman*). The *buddhi* is the container that stores all karmic impressions (*samksaras*) and transmutes across embodiments. These *samksaras* determine the conditions of one's next birth based upon actions that have come to karmic ripeness. The *antaḥkaraṇa* is the medium upon which the impressions left by the attention cast onto the manifest world by the *Ātman*'s *cit* are experienced as *vṛttis* (mental states). The illusion of separation is created by the *ahaṃkāra* (I-sense, imperfectly analogous to the Western concept of the ego).[56] Finally, the *citta* stores memories of the *vṛttis* as they arise from the grossest layers of the mind: the *manas* (thinking capacity for sense awareness with no external cause such as pleasure and thoughts) and the *indriya* (sense organs that receive perceptual input from the outside world).

10.5.2 Early Buddhism: Material reality and non-self

Buddhism is another vast and diverse set of traditions; this chapter focuses primarily on the Pāli canon of early Buddhism as promoted in the Pāli Abhidhamma movement and retained in the Theravāda school.[57] These early Buddhist teachings take an extreme opposite stance to Advaita Vedanta; while both share a degree of monism, early Buddhism places material reality as the starting point.[58] This, at least on its face, makes such trends in Buddhist thought more compatible with modern Western scientific approaches. Buddhism diverges from Western science with its belief in transmigration and karma.

The ultimate goal of Buddhist practice, *nirvana*, is the realization of the deeper truths of *anicca* (impermanence), *dukkha* (all physical embodied experience, no matter how pleasant, is inherently unsatisfactory) and *anattā* (there is no self). Because the underlying substrate of reality is ever-changing *kalapas*, everything is ephemeral and changing. Because everything changes, everything will ultimately end, leading to dissatisfaction (*dukkha*). Once one is finally enlightened to this fact of ultimate dissatisfaction, one is free to release complete and utter attachment to the illusion of a self and enter into the state of void no-thing-ness. Buddhism therefore posits an ultimate state of expanded awareness that emerges as self-dissolution, in contrast to the hyper-expansion of Self in Advaita Vedanta.[59]

The mind in early Buddhism starts from the particles of material reality (*kalapas* or bundles of the eight elements of reality) that build up to the *rūpī* (form) of the *kayo* (body[60]). The Buddha routinely refused to answer questions about how the eternal essence (Self/soul) was related to the physical form of the body, as he rejected the notion of an eternal self. Despite this materialism, Buddhists do retain belief in a life principle that transmutes across lifetimes and travels in and out of a body when dreaming.[61] While the interaction of an *ānanda* (consciousness) with a womb is the initial cause of a particular *nāma-rūpa* (mind–body) being formed, it remains an ultimately material phenomenon as this consciousness is bound to and an emergent property of the body. The *rūpa* generates the sense organs, which in turn generate sense experience, which gives rise to the *nāma*. Here the formal mind-body theory in early Buddhism comes close to recapitulating Cartesian dualism, in seeing the *nāma* and *rūpa* as distinct entities that, like a blind person (*rūpa*) carrying an immobile but sighted person (*nāma*), nevertheless require each other to cooperate for survival. Like modern neuroscientific theories of perceptual consciousness,[62] early Buddhist thought saw consciousness as arising from sense perception in combination with the memories of previous states stored within the *citta* (memory). The mind, and its emergent property of consciousness, is thought to be rooted in the body via a 'heart-basis' (*hadaya-vatthu*), which may be considered a rough parallel for the modern neuroscientific brain basis for consciousness, but here rooted in the heart rather than the brain. This heart basis is the physical support for the more subtle actions of the *mano-dhatu* (mind-sense, taken to be an additional sensory modality). The mind-sense enables *manasikāra* (attention) and *viññāṇa* (consciousness) to arise. It is this mind-sense that retains the *sankaras* (karmic impressions) that carry forward across cycles of reincarnation.

10.5.3 Sāṃkhya and classical yoga: Dual material/consciousness realities and distilled self

Sāṃkhya philosophy, and the associated practices of Patanjali's Rāja yoga, form a third primary lineage of thought about body and mind within Indian philosophy.[63] In contrast to the extreme idealist monism of Advaita Vedanta and the extreme material monism of early Buddhism, Sāṃkhya holds a strong dualist view that posits both consciousness (*puruṣa*) and matter (*prakṛti*) are equal, non-overlapping aspects that both operate to form the universe. Unlike Cartesian dualism, however, the mind is considered a part of material *prakṛti*. Like Advaita Vedanta, the material mind is a vessel through which the nonphysical, eternal, unchanging Self *Ātman* interacts with the physical *prakṛtic* world.

Sāṃkhya metaphysics posits that the material, manifested world occurs as a result of the *puruṣa*, which is constant, eternal, contentless and unchanging, observing the *prakṛti* that then activates from unmanifest to manifest changing, ephemeral, material reality. This point of interaction between *puruṣa* and *prakṛti* is *samyoga* (union, interaction) that leads to the continued *saṃsāra* cycles of reincarnation. The ultimate goal then becomes achieving *kaivalya* ('separatedness, aloneness') that eliminates this *saṃyoga* (conjunction), leaving the *puruṣa* to abide alone in its own completeness, completely disinterested in the *prakṛti*. When *puruṣa* stops observing *prakṛti*, the *prakṛti* stops moving and remains latent/unmanifest. In contrast to Advaita Vedanta, individual *puruṣa*s are discrete entities that do not ultimately unify into a single overarching consciousness.[64]

Sāṃkhya posits that mind emanates into material reality from this interaction with non-material consciousness in ways that are broadly similar to Advaita Vedanta, with the distinction that the *Ātman*/*puruṣa* is separate but equal to material *prakṛti* and each *puruṣa* remains wholly distinct from all other *puruṣas*. The mind is hierarchically arranged from more subtle to more gross: the *buddhi* (intellect) houses the *samskara* (imprints) of karma and transmutes across embodiments, the *ahaṃkāra* (I-sense) creates a sense of separate self that enables embodied experience, and the *citta* (memory) holds the records of previous interactions with sense objects. The sense modalities feed back into the *citta* and lead to the manifested versions of external objects by acting as the doorways through which the *puruṣa* creates manifested material forms in the *prakṛti* via observation.[65]

10.6 Other informal theories of mind

While formal study of the mind is not restricted to Western science or Indian philosophy, the vast majority of societies and daily interactions rely far more heavily on the informal models of minds that people use to interpret their social world. Though informal/folk models of minds can be as diverse as the cultures that hold them, this chapter focuses on two examples: animism associated with Indigenous societies' Traditional Ecological Knowledge systems and belief that minds are not knowable (or opaque) found throughout the Pacific.

10.6.1 Non-human minds: Traditional Ecological Knowledge and animism/perspectivism

Indigenous societies around the world tend to adopt a stance toward the natural world that sees all entities as having some unseen 'spirit' dimension that brings them all into relationship with each other.[66] This stance, called animism, has long been considered by Western researchers as a primitive, inferior epistemology,[67] though more recent work on the interplay between these beliefs and the extensive ecological knowledge Indigenous peoples hold alongside them reveals the sophistication of these world views.[68]

Amerindian perspectivism arose out of ethnographic work focusing on the cosmologies and ontologies of Amazonian peoples.[69] Taking some similarities to Vedic Indian philosophies, these cosmologies suggest that all beings began in an undifferentiated pre-self form. Though Amazonian perspectivism does not appear to hold belief in reincarnation, it does (like the Vedic Indian view) suggest that the spirit form is the original substrate of reality rather than material. Within Amazonian mythologies, the natural world and the cultural world of humans gradually diverged. While the West would place humans as growing up and out of nature, Amerindian perspectivism posits that all beings started out as humans and animals gradually lost the capabilities that humans retain.[70] This leads to a more relational attitude toward nature, as all beings are seen as having the same underlying consciousness as humans, merely hidden within different forms. The apparent underlying personhood of these entities are made visible through shamanic practices that allow the practitioner to look past the outward form. Though Western epistemology starts from a material realist stance and seeks to fully understand an entity or item from the outside looking in (objectivist), Amazonian perspectivism instead starts from the stance of spiritual realism and seeks to *subjectify* beings and even artefacts in order to comprehend them.[71] Anything which is capable of having a perspective has a subjectivity; the goal of understanding is to grasp what that subjectivity is.

Whether this animism and perspectivism is equivalent to the sort of promiscuous anthropomorphism suggested by some cognitive science of religion theories,[72] however, is not necessarily supported. These perspectives appear to indicate a more interconnected view of ontology, such that entities are ascribed properties in relation to how the subjectivity viewing them sees them. For example, a vulture sees the maggots on a rotting bit of flesh in the same way a human sees grilled fish, while a dead spirit sees boiled cassava as rotten meat.[73] Thus, food is food because of how it is related to the observer, rather than based upon some intrinsic characteristic. Developmental and cognitive anthropological work further suggests that young children in societies with these more integrated, relational natural ontologies do not over-extend the human mind in the way that Westerners do.[74] To the contrary, it appears that people in these societies learn from a young age to categorize knowledge about social agents in the world to include more diverse kinds of minds than the anthropocentric view held in Western perspectives.[75] The key socio-ecological factors that appear to promote these perspectives on minds include extensive personal experience interacting with the natural world and living within a society that is heavily focused on kinship to structure (human) social life.[76]

10.6.2 Opacity of mind: Minds are fundamentally unknowable

Anthropologists – especially those working in and around the Pacific – have challenged the notion that all people reference inner mental states as the default mode of social interaction.[77] Ethnographic works with small-scale societies in Papua New Guinea, Samoa, Fiji, in parts of the Arctic and amongst some Maya groups report participants consistently claim it is impossible to truly know someone else's mind. The mind, being ensconced within the opaque container of the body, is forever hidden from the outside world. These claims have been broadly described as the Opacity Doctrine or Opacity of Mind – here referred to as Opacity norms. These responses may indicate genuine differences in how people approach the problem of other minds in these societies,[78] or they may simply result from ethnographers being trapped within their own Western cultural models of mind and unable to properly formulate questions to fit their respondents' world views.[79] The most likely solution is some combination of these two, with consistent elements of social cognition expressed early in life across societies being modulated by cultural context across development.[80] While most adults in these communities will readily agree that other people have different thoughts, beliefs, desires, goals and knowledge, they often deny another person can know what these contents are. Gossip, however, remains pervasive and is often used to opine about the unseen intentions and desires of another; indicative both of ability to infer these states and the general normative proscription against doing so.

Opacity norms shift moral reasoning toward focus outcomes rather than intentions.[81] The Western model of mind holds that intentions are the best predictor of behaviour, leading attempted murder (a bad intention with a good outcome) to be seen as worse and more worthy of punishment than an accident (a good intention with a bad outcome). Here the violation is the thought (the bad intention) itself. Opacity norms, on the other hand, emphasize focus on the bad outcome of the accident; here the violation is the action, which is seen as more important than intent. Intent is still used to judge action, however, suggesting intent reasoning has some universality. Despite this apparent universality, intention in particular (and mind focus more generally) is used differently and developed differently in Opacity norm contexts.

The socio-ecological contexts that may drive development and retention of Opacity norms likely also relate to the special conditions of being on an island. As discussed with the Western model of mind, high degrees of autonomy and loose enforcement of social norms make minds a rational and efficient target for social understanding. In collectivist settings more generally, norms are more constraining on behaviour and therefore minds are less effective explanations for actions than the situation. However, this does not explain the lack of Opacity norms in most collectivistic societies; an additional boundary of even more restricted relational mobility (or the ability to move to a new social group) may have further provided the context for these norms to become particularly prominent in Pacific Island societies.[82] Reliance on kinship ties to structure daily life is another possible pressure promoting these norms, as both relational (im)mobility and kinship intensity reduce reliance on mental states via intentionality reasoning.[83]

10.7 Minds of humans and gods when cultures intertwine

Given the extent of colonization, missionization and globalization, it is impossible to clearly distinguish religious traditions in small-scale societies from the influences of the so-called world religions. If religious believers across traditions apply their own models of minds to supernatural agents, then one might predict that exploration into the expectations about what these supernatural agents are like will carry the social cognitive marks of the model of mind present within the culture where that religious tradition was developed.[84] Further, as religions spread through various forms of intercultural contact, the social cognitive signatures of their originating cultural model of mind may be adopted along with the new religious belief sets.

Despite the ravages of European colonization and Spanish Catholic missionization in the Americas, Indigenous beliefs that incorporate connections to land and perspectivism continue on.[85] In the first centuries of European occupation, Mayan communities took on the necessary trappings of Catholicism to appease their overlords, but retained their traditional beliefs and practices both in secret and more openly in the case of overlapping elements (e.g. the cross as a symbol sacred to both Maya and Christianity).[86] As Indigenous beliefs were gradually given more space to be openly expressed and elements of introduced beliefs were integrated, uniquely Amerindian interpretations and focuses on beliefs spread. For example, as Amerindian communities are often more focused on communal life in the here-and-now, Christian teachings about afterlife salvation are diminished while beliefs in Jesus as a healer are promoted.[87] At the same time, Christian imagery and concepts can be invoked in traditional spiritual and medicinal modalities (e.g. ayahuasca ceremonies[88]).

As indicated earlier, Protestant Christianity stands out as one of the most mind-focused religious belief sets. Unlike Jewish or Roman Catholic believers, Protestants are more likely to agree that the mere thought of wrongdoing constitutes sin.[89] Missionization has spread Christianity around the world; Protestantism has gained particular prominence in Opacity of Mind societies around the Pacific. Ethnographers have noted distinct rises in distress associated with introduced Protestant practices that go against the dictates of private mental lives within Opacity norms.[90] Practices like confession compel believers to speak openly, sincerely and (worst of all, publicly) about their innermost states. The resulting loss of mental privacy and freedom leads to increases in feelings of shame and spikes in damaging gossip.[91] Though these societies had pre-missionization practices for adjudicating interpersonal grievances, they often retained the privacy of one's inner world. For example, Fijians have ritualized practices like *i soro* (surrender) that are used to repair relationships in cases of wrongdoing. Conducting an *i soro* involves a public admission of one's actions and a transfer of goods from the transgressor to the aggrieved parties (and/or their kin). However, these *i soro* events are engaged in on a voluntary basis and do not delve into the inner motivations of the wrongdoers as Protestant confession practices dictate.[92] Though Christianity is often considered a core part of identity in Pacific Island communities,[93] Christian God concepts can still be seen to support local understandings of minds and behaviours.[94]

10.8 Conclusion

Though the Western understanding of the world as inhabited by autonomous beings who act based upon inner states, instantiated in the brain, has led to massive growth in our understanding, it is only one of many views on the relationships amongst the brain, body, mind and behaviour. Other formal theories of mind like those found in Indian philosophy and informal theories of mind like the animism/perspectivism of the Americas and the Doctrine of Opacity in the Pacific show us that our abilities to perceive and conceive of the mind are heavily influenced by the cultural settings we inhabit.

Notes

1 Lillard 1998.
2 Curtin et al. 2020; Krull et al. 1999; McNamara et al. 2019.
3 Barrett et al. 2013; Callaghan et al. 2005, 2011; Duranti 2015; Mayer and Trauble 2012.

4 See for example Chakroff and Young 2015; Frith 1999; Gray, Young and Waytz 2012; Malle 2006.
5 D'Andrade 1987; Spunt and Adolphs 2015; Wierzbicka 2006.
6 Markus and Kitayama 1991; Taylor 1989.
7 Bloom 2004; Chudek et al. 2017; Starmans and Bloom 2012.
8 Kelemen 1999; Malle and Knobe 1997; McNamara et al. 2019, 2021; Mull and Evans 2010; Woodward 2009.
9 Ross 1977.
10 Inglehart 1997; Norris and Inglehart 2004.
11 McNamara et al. 2019, 2021; Walker, Smith and Vul 2015.
12 Henrich 2020; Taylor 1989, 2007.
13 Laine 2014.
14 Becker et al. 2017; Choe, Keil and Bloom 2012; Chudek et al. 2017; Hood, Gjersoe and Bloom 2012; Seth et al. 2006; Starmans and Bloom 2012.
15 Azari et al. 2001; Bareham et al. 2020; Kalbe et al. 2010; Lebedev et al. 2015; Millière 2017; Mormann and Koch 2007.
16 Miller 2003.
17 Andersen 2017; Clark 2013; Powers, Kelley and Corlett 2016; Ridder, Verplaetse and Vanneste 2013; Seth, Baars and Edelman 2005.
18 D'Andrade 1987; Heyes and Frith 2014; Malle 2006; Povinelli and Giambrone 2001.
19 515.
20 Brune 2006; Frith and Frith 2006; Leslie 2001.
21 Titchener 1924, 417.
22 Baron-Cohen and Wheelwright 2004; Davis 1983; Davis 1980; Fitzgibbon et al. 2010; Sebastian et al. 2012.
23 Batson 1997; Cialdini et al. 1997; Waal 2008.
24 Schaafsma et al. 2015.
25 Baron-Cohen 1995; Frith, Morton and Leslie 1991.
26 Ali and Chamorro-Premuzic 2010; Blair 2008; Mullins-Nelson, Salekin and Leistico 2006.
27 Mikolajczak et al. 2007; Moriguchi et al. 2006.
28 Gray, Gray and Wegner 2007; Gray, Young and Waytz 2012.
29 Barrett 2011; Boyer 2001; Guthrie 1993.
30 Johnson 2009; Norenzayan et al. 2016; Schloss and Murray 2011; Sosis 2009.
31 See for example: Barrett 2011; Barrett 2000; Barrett and Keil 1996; Bering 2006; Boyer 2001; Guthrie 1993.
32 Bering 2006.
33 Guthrie 1993.
34 Gervais 2013; Norenzayan, Gervais and Trzesniewski 2012.
35 Caldwell-Harris et al. 2011; Genovese 2005; Gray et al. 2011; Willard and Norenzayan 2017.
36 Epley et al. 2009; Kapogiannis et al. 2009; Schjoedt et al. 2009.
37 See, for example, Genovese 2005; Riekki, Lindeman and Raij 2014; Willard and Norenzayan 2013, 2017.
38 Gervais and Henrich 2010; Luhrmann et al. 2021; McNamara et al. 2021.
39 Cohen and Hill 2007; Graham and Haidt 2010; Laurin, Kay and Fitzsimons 2012.
40 Airini et al. 2010; Brison 2007; Carsten 1995; Groark 2011; Heine 2001; Kavapalu 1995; Markus and Kitayama 1991.
41 Power distance: Hofstede 1983.

42 Gelfand, Nishii and Raver 2006; Leung and Cohen 2011; Vandello, Cohen and Ransom 2008.
43 Matsumoto 1989; Matsumoto, Yoo and Fontaine 2008.
44 Jha 2017; Sriraman and Benesch 2005.
45 Asad 2003; Weiskopf 2020.
46 See for example: Sriraman and Benesch 2005.
47 Berry 2015; Henrich, Heine and Norenzayan 2010; Nielsen et al. 2017; Rad, Martingano and Ginges 2018.
48 Keyes and Daniel 1983.
49 Markus and Kitayama 1991
50 Bodewitz 2019; Bryant 2015.
51 Peacock 2008; Whicher 1998.
52 See Loy 1982.
53 Adams and Indich 1982; Schweizer 2020; Vilela 2019; Weed 2006.
54 Sriraman and Benesch 2005.
55 Bilimoria 1980; Mayeda 1992.
56 Tayal and Sharma 2020.
57 Harvey 1993; Loy 1982; Osto 2018; Samuel 2014.
58 Prasad 2000; Samuel 2014.
59 Loy 1982.
60 See: Harvey 1993; Loy 1982; Ronkin 2005.
61 Harvey 1993; Prasad 2000.
62 Bareham et al. 2020.
63 Bryant 2015; Osto 2018; Schweizer 1993, 2019; Whicher 1998.
64 Bryant 2015; Virupakshananda 2015.
65 Schweizer 1993, 2019; Whicher 1997, 1998.
66 Bird-David 1999; Hall 2011; mays, Seligman and Medin 2020; Rival 2012.
67 Bird-David 1999.
68 Berkes, Colding and Folke 2000; Smith 1998; Weiskopf 2020.
69 Callicott 2013; Castro 1998, 2004; Rival 2012; Turner 2009.
70 Castro 2004.
71 Castro 1998, 2004.
72 Guthrie 1993.
73 Castro 2004.
74 Herrmann, Waxman and Medin 2010; Medin et al. 2010; Unsworth et al. 2012.
75 Mays, Seligman and Medin 2020; ojalehto, Medin and García 2017a, 2017b.
76 Bird-David 1999; Castro 2004; Hall 2011; Toyoda 1998.
77 Duranti 2015; Hollan 2012; Luhrmann 2011; Throop 2012; Toren 2009.
78 Robbins 2008; Robbins and Rumsey 2008.
79 Duranti 2008; Groark 2011.
80 Curtin et al. 2020; Dixson et al. 2017; Heyes and Frith 2014; McNamara 2016.
81 H. C. Barrett et al. 2016; Curtin et al. 2020; McNamara et al. 2019.
82 McNamara et al. 2019, 2021.
83 Curtin et al. 2020; McNamara and Henrich 2017; Thomson et al. 2018.
84 McNamara et al. 2021; McNamara and Purzycki 2020; Purzycki 2013; Purzycki et al. 2012; Purzycki and McNamara 2016; Purzycki and Sosis 2011; Willard and McNamara 2019.
85 Cook 1997; Kidwell, Noley and Tinker 2020; Radding 1998; Watanabe 1990.
86 Cook 1997.

87 Kidwell, Noley and Tinker 2020; Radding 1998; Watanabe 1990.
88 Luna 1986; Wright 2018.
89 Cohen and Hill 2007.
90 Duranti 2008; Keane 2008; Robbins 2008; Robbins, Schieffelin and Vilaça 2014; Schieffelin 2008.
91 Duranti 2008; Keane 2008; Robbins 2008; Robbins, Schieffelin and Vilaça 2014; Schieffelin 2008.
92 Arno 1976.
93 Ryle 2010; Tomlinson 2007; Toren 2004.
94 McNamara et al. 2021; Willard and McNamara 2019.

11

On religious practices as multi-scale active inference

Certainties emerging from recurrent interactions within and across individuals and groups

Inês Hipólito and Casper Hesp

11.1 Introduction

Religious practices are collective human activities under a cultural setting,[1] centred on co-constructing spiritual meanings in rituals and sacred histories,[2] thus requiring a variety of cognitive capacities.[3] Personal faith experiences tend to involve perceived contact with something sacred, combined with a perceived transcendence that provides meaning and direction for life.[4]

While awareness of religious diversity is not new, philosophers of religion from both East and West have urged for more interactive collaborations.[5] There are both striking similarities and differences between the approaches put forward within the most popular religious traditions across the globe. While we cannot do justice to all their nuances, it can be insightful to highlight some of their key features in the following order: self-relation, salvation, devotion and veneration.

First, religious traditions propagate different ways of relating to ourselves, as contextualized by one or more larger realms. Buddhists assert there is only Oneness (strict monism), which inspires the common Buddhistic claim that the notion of an individual 'self' is fictional in essence. Hindus commonly wield the concepts of *Brahman* (connoting divine ultimate reality) and *Ātman* (connoting the substance of individual consciousness or 'self'), but various Hindu communities propagate divergent views on whether this distinction is essential or conceptual, affording both dualistic and non-dualistic views. Non-dualistic Hindus tend to assert that 'Ātman is Brahman', which implies that everyone is permeated by divine consciousness (e.g. Advaita Vedanta Hinduism). In contrast, adherents of the Abrahamic religions tend to believe in an almighty Creator that differs from creation. For most Jews and Muslims this distinction between God and mankind is absolute in every substantial sense. In contrast, most modern Christian traditions believe in the Trinity (Nicene Creed,

325 AD), which claims that God ('the Father') incarnated fully in Jesus Christ ('the Son') who enabled his followers to be filled with God's spirit ('the Holy Spirit').

Second, a central theme in the world's religions is to assert a spiritual yearning of humankind and offer some resolution, often referred to as 'salvation'. Within the Abrahamic religions, salvation typically involves being accepted by or united to God while retaining one's individuality, although traditions differ widely on how one attains salvation. A similar theme is central for dualist Hindus, who strive for unity with Brahman while differing on best practices. For non-dualistic Hindus, this unity is already there in essence, but it is 'hidden' until individuals achieve *moksha* (liberating self-knowledge). While there are similarities, Buddhists deny the existence of the self more explicitly – their ultimate goal of *nirvana* involves a complete extinguishing of the self and extinction of all suffering.[6]

Third, although practices are widely divergent across religions, all involve some form of personal devotion to a larger whole that is perceived as sacred, which usually includes an overarching community and its established authorities. In daily life, such devotion tends to be expressed and maintained in many ways, both publicly (e.g. through clothing, generosity and other observable behaviours) and privately (e.g. in one's thought). On the individual level, there is an ongoing, circular relationship between one's personal devotion or 'faith' and one's expressions thereof. Devotion tends to be deepened by expressions thereof, such as community involvement and meditation/prayer, which are practised both individually and collectively.

Fourth, the practice of veneration has been identified as an integral part of each of the major world religions, either explicitly or implicitly. Highly dedicated devotees of religious communities can be regarded as sacred themselves ('saints') and respected as authorities on religious matters (e.g. morality). More implicitly, community-level perceptions of a community member's degree of devotion will tend to strengthen the perceived trust in that person's authority on religious practices. At the same time, religious devotion encourages submission to religious authorities. Therefore, due to (implicit) veneration, the relationship between religious devotion and religious authority is intrinsically circular, cumulative and collective.

Next, we will survey some discussions in the cognitive science of religion. For example, the form of religious beliefs themselves has been a subject of debate[7] as scholars have tried to pin down the determining conditions for religious convictions, and how they relate to evidence-based reasoning and/or 'blind' faith.[8]

Cognitivist approaches to religion tend to focus on religious beliefs and regard them as cognitive processes of (implicit) computational inference[9] and some have argued they should be seen as misguided 'alternatives to empirical reality'. Specifically, Wood and Shaver (2018) state that 'the evolutionary roots and continued persistence of religion are thus functions of the capacity for humans to generate cognitive alternatives to empirical reality, and the need to stably coordinate those alternative conceptions'[10] and cast religious beliefs as unnecessary cognitive processes that require correction. If religion could indeed be reduced to claims that need to be tested scientifically, one could paint a picture of human progress away from 'erroneous' magical and religious beliefs, and towards scientifically validated thought, as described by Frazer (1890) and Dennett (2007).

While the scientific method is (and should remain) dedicated to producing natural explanations for all observable phenomena, including religion, many authors have criticized various parts of the overall perspective of human progress put forward by Frazer (1890) and Dennett (2007), for example, by arguing that enculturated practices in general require contextualization and cannot be properly understood by reduction to cognitive processes alone.[11] Reductive cognitivism ends up disregarding the diverse kinds of emergent dynamics entailed in religious practices both within and between individuals and communities over time, such as the roles played by non-propositional meanings, cultural embodiment and the broader sociocultural environments in which religious practices are situated.[12]

In this chapter, we develop the implications of viewing religion as an enculturated practice and evaluate it in light of cognitive science and philosophy of mind, as inspired by the consonance between Wittgenstein's thought and the conceptual toolbox associated with complex adaptive systems. In doing so, we converge on a non-reductive, communal perspective on religion that contrasts starkly with common 'straw person' perspectives on religion that aim to reduce it to 'erroneous' metaphysical theorizing motivated by spiritual experiences 'generated by the brain'. In contrast, a communal perspective informed by Wittgenstein's thought, cognitive science, and complex systems theory suggests that religious practices both enable and require the emergence of ineffable meanings that are experienced both individually (i.e. one's personal faith) and collectively ('the faith'), with only partial commensurability. Individual participation in collective religious practices can generate evidence that becomes commensurable for other encultured participants, as personal experiences that support their individual faith in the collective and consolidate future participation. In this view, religions emerge from a recursive relationship between individual and collective dynamics that cannot be reduced to the individual level of description. We further elaborate on the collective and perspectival aspects of religion that are incommensurable with the scientific method. While imposing such an epistemic separation has proven useful in practice, we furthermore argue that a harsh dichotomy is itself an oversimplification that does not provide a 'perspicuous presentation' of science–religion interactions. Touchpoints between these different kinds of language-games generate perpetual confusion and unproductive discussions. As a particularly impactful example, we discuss ethical considerations in policy-making, which demonstrates how our account sheds light on complex societal challenges at the science–religion interface.

11.2 On reductive cognitivism in the study of religion

Religious practices require participants to master a diverse set of cognitive skills. While situated within and enacting a specific sociocultural setting, religious individuals actively engage in ritualistic activities, such as festivals, feasts or meditation. In doing so, they engage with sacred narratives that shape their lives and their relationships with others in ways that tend to be experienced as meaningful and purposeful.

Much of the (neuro)cognitive science of religion tends to focus on the cognitive processes associated with religious beliefs. One of the most prominent reductive views on religion is championed by Pascal Boyer (2001; 2018). In this view, religious concepts and beliefs are cast as unconscious inferences computed in modular systems, following the Modularity of the Mind framework (hereafter MM[13]). This framework contends that cognitive processes take place in a roughly double-layered architecture. On the bottom levels, modular mechanisms are encapsulated to generate (series of) responses given some set of domain-specific stimuli. The resulting responses are continuously aggregated in domain-specific representations that feed upwards to the central module, which is tasked with integrating its various uni-modal inputs in one centralized 'mental representation'. According to cognitive reductivists, 'religious beliefs' are formed at this level of MM. Boyer provides the following examples of this process:

> When we see branches moving in a tree or when we hear an unexpected sound behind us, we immediately infer that some agent is the cause of this salient event. We can do that without any specific description of what the agent actually is.[14]

In MM, certain series of stimuli will trigger simplified perceptual representations in the corresponding domain-specific modules (e.g. visual or auditory) and those are fed to the central cognitive module, which then infers what appears to be the most likely cause of this perceptual representation (e.g. another agent). This is a reasonable example scenario where unexpected sensorial stimuli are attributed to another agent, supposedly hidden. In Boyer's account, 'Specialized explanatory devices, more properly called inference systems, each of which is adapted to particular kinds of events and automatically suggests explanations for these events. . . . Religious concepts may seem out of the ordinary, but they too make use of the inference systems I just described.'[15]

Although formal reasoning about the existence of other minds is challenging,[16] it sidesteps much more complicated questions such as spiritual transcendence and other aspects associated with religious experiences. It leaves unanswered important questions about how a cognitive architecture that gradually develops representations in an entirely stimulus-driven fashion is expected to construct representations associated with transcendent spiritual realms and entities. Postulating that neural architectures employ symbolic representations remains highly controversial.[17] Further, suggesting that our capacity to make inferences about abstract concepts that commonly recur in our daily lives could have been repurposed and generalized to generate hypotheses about spiritual beings and their intentions does not clarify how that qualitative transition occurs for the individual or the collective, nor does it shed any light on its practical implications for the daily lives of those who entertain and propagate spiritual concepts. In short, MM does not provide clarity on the formation and propagation of spiritual concepts and religious beliefs more generally.

Contemporary cognitive empirical and theoretical research has generated multiple lines of evidence contradicting MM.[18] Some researchers specifically note a Kuhnian shift in the neurocognitive science of religion,[19] away from MM (mostly bottom–up) towards predictive accounts of cognition that emphasize and formalize recursive relationships in which the relative dominance of bottom–up and top–down influences varies dynamically.[20] Predictive accounts of cognition have become predominant as explanations of cognitive processes based on a large amount of scientific evidence for the role of domain-general over domain-specific processing, disputing the information encapsulation supposed by MM.[21] Another challenge for MM is the accumulating evidence suggesting top–down effects on lower-level cognition, which has led to predictive coding becoming the basis for a variety of frameworks to understand neural activity.[22] Andersen (2019) specifically applies predictive coding to the detection of supernatural agency. He argues that most false positives in agency detection can be seen as top–down interference of overweighting unreliable stimuli. In his view, the formation of religious beliefs is a consequence of an error of the predictive system, which, in the face of unreliable stimuli, puts too much weight on priors.

Reducing religious practices to religious beliefs, understood as inference of a mental representation or as an inferred mental representation, under predictive accounts of cognition, motivates thinking of religious practices as inference alternatives to the empirical reality.[23] This motivates the consideration that those who engage in religious practices are somewhat under a predictively faulty mechanism, reasoning or model that needs to be updated for epistemic accuracy. In this light, predictive accounts of cognition also do not elucidate upon the belief/concept formation about transcendent beings and what these concepts mean for human life, other than being 'fictional'. This explanation depends upon accepting the following conditions: (1) evidence is only admissible if it is found by explicit reasoning and (2) cognitive capacities implied in religious practices can be reduced to such explicit reasoning (i.e. it is sufficient to explain subjective experience, of the ethical, the religious or the psychological). Accordingly, such predictive accounts of cognition tend to conclude that inferences about something that is not in the empirical world must be corrected for epistemic accuracy.

The intention to inspire progress from religious towards scientific thinking is also held by Dennett (2006). In his popular book *Breaking the Spell*, taking inspiration from Frazer's (1890) analogy between religion and magic, Dennett argues that the phenomenon of religion requires scientific analysis, for it must not be off-limits to or beyond scientific inquiry. In his view, science should study religion by applying evolutionary biology and memetics to beliefs in a supernatural agent and consider the implications for the society of individuals being under a certain spell or error of reasoning.

In contrast, Wittgenstein[24] disputed the view of religious practices as an error of reasoning that must be corrected: 'Frazer's account of the magical and religious views of [humankind] is unsatisfactory: it makes these views look like errors.'[25] He continues: 'it is nonsense for one to go on to say that the characteristic feature of these actions is the fact that they arise from faulty views about the physics of things.'[26] On the contrary, nonsensicality itself is celebrated in certain religious doctrines, based on the belief that valuable truths lie beyond the capacities of explicit human reason. For example, the two-truths doctrine of Buddhism encourages its practitioners to simultaneously entertain multiple statements that are mutually exclusive, motivating Catuşkoṭi, an intrinsically paradoxical (or 'paraconsistent') logic system. Furthermore, paradoxical statements are central to Christian doctrines, for example as discussed by Søren Kierkegaard, whose work 'Practice in Christianity' celebrated Christ as the 'Absolute Paradox', a perspective that was further popularized by theologian Karl Barth.[27]

In such cases, religious statements appear nonsensical because they are attempts to move beyond our everyday realities and hence beyond the significant language.

The human mind is often lured into going against the boundaries of language. In this regard, Diamond (2019) made a distinction between what is knowable and what is sayable, precisely to demarcate that some things that can be known are not sayable, which she terms 'mystical'. Because certain core parts of religious convictions are not in the domain of the sayable, those parts are by definition not contradictable:

> And then I give the explanation: 'I don't believe in . . .', but then the religious person never believes what I describe.
>
> I can't say. I can't contradict that person.[28]

This thought is evident in the train of propositions leading to the famous last proposition in the Tractatus, '7. Whereof one cannot speak, thereof one must be silent'. Wittgenstein refers to the kind of mystical knowledge concerning the problems of life, for example ethical or religious, that is known but not sayable.

In Wittgenstein's view, many philosophers are driven by a 'craving of generality' and 'preoccupation with the method of science' that motivate reductive tendencies.[29] Naturally, cognitive models and experimental research are important means for understanding neurocognitive processes underlying and enabling human sociocultural practices in general and religions in particular. What is disputed is whether the method of science is appropriate everywhere and specifically in religious considerations. In the case of understanding religious practices associated with, for example, spiritual or mystical experiences with ineffable properties,[30] it is not possible to set up a research experiment to answer this question. In this respect, Wittgenstein remarks the following: 'I believe that the attempt to explain [religion] is already therefore wrong, because one must only correctly piece together what one knows, without adding anything.'[31] Religious practices are deeply rooted in the experience of the mystical, hence it is not possible to explain religion without adding how it is in relation to me. Indeed, 'Here one can only *describe* and say: this is what human life is like. Compared with the impression which the thing described makes on us, the explanation is too uncertain. Every explanation is after all a hypothesis.'[32] This shows, Wittgenstein suggests, 'that it is absurd to say "Science has proved that there are no miracles"'.[33]

In conclusion, religious considerations with regards to human living are not comparable to scientific hypotheses. Only in theorizing can there be error. According to Wittgenstein, neither Augustine nor the Buddhists were in error, 'except when they set forth a theory'.[34] Paraphrasing Wittgenstein, there is nothing hypothetical in our religious considerations.[35] For such a reason,'[a]ll explanation must disappear, and description alone must take its place'.[36] Religious practice emerges from a certainty only describable by stating 'how things are [to us]'.[37]

11.3 On the certainty of religion in enculturated practices

In this section, we expand on Wittgenstein's notion of certainty as a practice by employing the famous rule-following paradox. While theology is a discipline par excellence, one participates in religious ceremonies as forms of acting that give expression to membership of a religious community. One engages in religious practices with some degree of certainty or confidence in one's religious community – with faith.

Certainty is, according to Wittgenstein, both an attitude and the object of an attitude, that is, things one does without a doubt.[38] Certainty is a basic disposition that is non-propositional yet unwavering. In other words, certainty is a kind of conviction that does not rely on propositions, but on acting:

> the end is not certain 'propositions' striking us immediately as true, i.e. it is not a kind of seeing on our part; it is our acting, which lies at the bottom of the language-game.[39]

Certainties exhibit themselves in the attuned and adjusted ways by which one navigates and co-constructs the world. Religious practices, like any other sociocultural practice, are a way of acting amongst many potential ones – they are facts of our living:

> the fact that we act in such-and-such ways, e.g., punish certain actions, establish the state of affairs thus and so, give orders, render accounts, describe colours, take an interest in others' feelings. What has to be accepted, the given – it might be said – are facts of living.[40]

'Facts of living' are hardly ever acknowledged either because 'of their great generality'[41] or because they are always before our eyes.[42] Yet, facts of living ground certainties, certainties themselves grounding the ways of our living, what Wittgenstein calls, 'forms of life':

> Now I would like to regard this certainty, not as something akin to hastiness or superficiality, but as a form of life [Lebensform].[43]
>
> But that means I want to conceive it as something that lies beyond being justified or unjustified; as it were, as something animal.[44]

Certainties via the forms of life are what is to be accepted as a given or that is beyond justification. As opposed to a form of thought, calculating, perceiving and engaging in a religious ritual is an animal attitude deployed in our ways of acting, resembling a form of living. In other terms, certainties are thereby the kind of animal attitude that deploys itself in the ways of our acting.[45]

How are beliefs – that are de facto held – manifested in our acting? Do beliefs instruct our acting or is it in our acting that we do our believing? We motivate the latter: it is in our acting that we do our believing. Rule-following, we claim, provides a way of living and is sufficient for understanding certainty and action as basic, non-propositional, non-epistemic and yet an indubitable attitude towards some things. Rule-following implies a number of intertwined issues in epistemology that are relevant for how we think of (religious) practices: what is a (religious) rule? Can a rule compel or guide its user on how to act, and if so, how? What is it to be committed to following a rule? Is it possible to understand rule-following without referring to communities?

Those who champion a belief-first epistemology[46] contend that true knowledge depends on representing a rule that expresses what must happen or what does happen when certain conditions are met. In this view, one knows what to do and how to act upon the world by representing guiding rules. What these diverse accounts of action representation – as far apart as the propositional[47] or the practical[48] – have in common is that they too are belief-driven.[49]

Wittgenstein's notion of rule-following presents a problem to belief-first epistemology as an explanation of (religious) practices. At stake is that, in order to represent a rule we need to enter an infinite regress of further rules, or an interpretation, that cannot be given by the rule itself. The paradox is posed in the *Philosophical Investigations* as follows:

> This was our paradox: no course of action could be determined by a rule, because every course of action can be made out to accord with the rule.[50]

Wittgenstein's paradox resonates with Greek scepticism in implying that absolute knowledge is impossible. There are no normative facts about how to correctly apply a rule (all we have are facts of living), so we cannot possibly know how to correctly apply any rule (if knowledge is factive). Kripke (1982) used simple arithmetic rules to illustrate that one cannot use past behaviour to justify present behaviour considering that nothing from past behaviours guarantees how to generalize the rule to novel situations. The finite number of past instances will always be consistent with infinitely many conceivable alternative rules. This insight is generalizable to other aspects of mental life and even faith: one's present usage is a 'leap in the dark', that cannot be fully proven based on past usage.

One might propose drawing on the explicit rule representations instead.[51] This would still not solve the problem, as interpreting such representations would require even more rules. Perhaps one should reduce rules to more basic rules, which would hopefully allow them to offer a non-circular definition and understanding of addition. Yet the sceptical problem remains. How does one know one is following a particular rule? To follow a rule is a form of proceeding amongst all possible manners of proceeding. Notably, whatever rule someone gives themselves, it cannot by itself determine the correct use of an associated symbol or expression. For any procedural result obtained by supposedly following a rule, there is a deviant or alternative interpretation to the rule, ad infinitum. A rule, regardless of its Mentalese symbols the rule is encoded with, would then always require some interpretation to determine its meaning. Even if someone examines their own mental life or even the belief in God, this belief would still require some interpretation to determine its meaning where no fact would be discovered.

Does this mean that meaning is interpretation? Surely not, because this would imply infinite regress. Understanding an action or an expression would require an interpretation that would not itself need an interpretation: one 'last interpretation'.[52] The issue is that 'every interpretation still hangs in the air along with what it interprets, and cannot give it any support. Interpretations by themselves do not determine meaning'.[53] It seems however that we succeed on a daily basis to generalize a variety of societal norms to novel situations. If we cannot do so purely by past experience, mentally representing the rule nor by interpreting it, then how do we ascertain what rules are being followed? This sceptical challenge generalizes to all beliefs that can be expressed by language: there are no facts about meaning in general.

Explaining a procedure through justifications would lead to an infinite regress. Applied to religious belief, one could attempt to explain why one has attended a certain religious event by providing the reason that one believes in a certain God. But then again, justifications do not stop here, they will go on ad infinitum.

Justifications must come to an end somewhere.[54] Where they do, there is no further rule. It is our acting that lies at the bottom of the justification-driven language-game.[55] The sceptical challenge is dissolved when one denies the sceptic's request for an extra-linguistic entity that dictates a standard for the correct application of the terms considered rules, that is an objective normativity. We could obey a rule in our practice: 'And hence "obeying a rule" is a practice.'[56]

In our practices we do our believing. Certainties are in our actions.[57] There is a 'misleading appearance that leads philosophers to believe that at the foundation of thought is yet more thought'[58] (Moyal-Sharrock 2016). Interpretation must end in the animal in epistemology. More specifically, 'hinge certainties' as the ungrounded, non-propositional underpinning of knowledge, manifest themselves as ways of acting. It is in religious practices that people enact beliefs about the ineffable. Those beliefs are grounded in the basic certainties of their forms of life. In the next section, we will expand on this notion of religious belief as a practice, we will argue that religious practices instantiate a form of rationality.

11.4 On language-games and embodied rationality in religious practices

In this section, we will be arguing that religious practices, considered in their own context, involve an embodied form of rationality. Because (P1) religious practices involve the embodiment of language-games; (P2) language-games are a form of life (consisting both in language and practice), and (P3) rationality is a matter of making allowed moves within language-games, (C.) religious practices involve a form of embodied rationality.

Wittgenstein's treatment of 'rationality' is a rule-governed grammar of the ordinary and everyday that gives expressive force to a form of life. This deflationary, context-dependent sense of rationality provides a stark contrast with notions of logical empiricism or positivism, adherents of which would tend to uphold science as the greatest embodiment of rationality.[59] In *Philosophical Investigations*, the two most striking features of language-games is that while (1) they embody the regularities of a sociocultural community, and (2) they are also subject to transformation as the communities change. The 'framework of the game' is subject to transformation over time reflecting changes in the form of life.[60] The crucial point is that rationality cannot be reduced to fixed meanings since the meaning is in-use change and thereby reflects the changes of rationality, rational reasoning and action.

Apart from the openness to transformation, language-games also embody regularities. They contain the combination of linguistic expression and action, that is to say, the governing rules of a language-game. A language-game is a matter of, not representing a rule, but rule-following. Following a rule, as seen in the previous section, is a matter of actions and reasoning about possible actions, which is par excellence a form of rationality. For this reason, rationality cannot be arbitrarily defined in human nature, but only in human language and action.[61] Wittgenstein's notion of language-game offers a perspective of rationality as 'form of life'. More precisely, a perspective of rationality that, while preserving the rules, is also sensitive to sociocultural situatedness. Rationality becomes part of an activity where '[i]t is the language-game that manifests a rational behaviour and not the person herself'.[62]

Language-games express facts about ethics,[63] religion,[64] aesthetics,[65] science[66] or any culturally situated practice. The crucial insight for the topic of our chapter is that religious practices, particularly, involve language-games. Language-games are a form of life; that is, they consist of both language and practice. Speaking a language is part of an activity or a form of life in a specific sociocultural community. In this respect, Wittgenstein remarks as follows:

> Children learn their native language and acquire the rules through learning to play language-games. Learning to play the game is understood in terms of making a move in the game. Learning to play the game is a paradigm of learning to be rational.[67]

Rationality is a matter of internal consistency, making allowed moves within language-games. Rationality implied in moving around a language-game cannot hence be reduced to fixed meaning. For the stability of meaning is usage, that is, action. Rationality corresponds to the family of concepts that are part of an activity: where participation in the language-game corresponds to rational behaviour.

From a Wittgensteinian perspective, religious practices involve a form of embodied rationality. The use of the concept 'God', according to Wittgenstein, expresses the rules for the use of theological terms in everyday religious discourse.[68] Religious discourse is a form of grammar that allows us to express, not 'scientific facts' (whatever that is supposed to mean), but how things are experienced by practitioners. This is evident in the following passage: '"You can't hear God speak to someone else, you can hear him only if you are being addressed" – That is a grammatical remark.'[69] The remark is grammatical[70] because the use of 'God' in the context of religious worship differs from everyday discourse in its deep grammar.[71] In some of his writings, Wittgenstein agreed with Luther in saying that 'theology is the grammar of the word "God"' (Lectures). Thus, standard claims of theology about God are not to be understood as scientific claims but rather 'as grammatical remarks expressing rules for the use of theological terms in everyday religious discourse'.[72] We can talk about such matters in a way that shows the ineffable, not as scientific facts but as grammar.

In this section we have argued that because religious practices involve language-games, which are forms of life (consisting both in language and practice), and rationality is a matter of making allowed moves in language-games, religious practices involve a form of embodied rationality.

11.5 Developing perspicuous presentations of religion in light of cognitive science and multi-scale generative modelling

The sheer diversity of the various angles we have considered in this chapter so far might be overwhelming to the reader. The crux of our argument entails that this diversity is absolutely necessary, should anyone wish to provide a perspicuous presentation of the complex, multi-scale phenomenon called religion, without particularly favouring any particular level of description.[73] The human mind's reductive tendencies will favour latching onto a particular level (e.g. genetic, neurodevelopmental, affective, cognitive, social) and push for the elimination of others by subsuming all their explanatory power under one's current favourite. The conceptual toolbox of complex systems theory[74] allows us to cast levels of description as providing fundamentally different, yet complementary perspectives on a complex phenomenon, while acknowledging that none of them individually provides us with the full picture. Throughout the chapter, we have developed the implications of viewing religions as enculturated practices, which requires a primary focus on believing as (en)acting. However, we also intend to avoid a complete reduction to action, because one's subjective experience of religious practices is of an intrinsically perspectival nature that cannot be reduced to one's actions alone.

Our multi-perspectival approach is most properly understood in light of multi-scale formulations of Markov blankets as developed by Friston (2019) and colleagues[75] to formalize how interactions between internal and external states of any particular system (or 'particle') are mediated by its 'blanket states'. Oftentimes, these blanket states can be further decomposed into sensory states (inward influences) and active states (outward influences). Importantly, the notion of 'state' and the internal–external distinction are both observer- (or model-)dependent and thus fully generalizable. This statistical abstraction allows us to consider different levels of description and translate between these perspectives, a kind of move that has special relevance to our discussion of religion. In particular, Friston's monograph (2019) demonstrates the prime importance of blanket states when moving to higher (i.e. slower, larger, more coarse-grained) levels of description:

> When lifting the dynamics from one scale to the next, internal states are effectively eliminated, leaving only slow, macroscopic dynamics of blanket states. These become the states of things at the next level, which have their own Markov blankets and so on. The endpoint of this formalism is a description of everything at progressively higher spatial and temporal scales. The implicit separation of temporal scales is used in subsequent sections to examine the sorts of dynamics, physics or mechanics of progressively larger things.[76]

In Figure 11.1, it is illustrated how the recursive application of reduction and partition operators affords analyses of a system's dynamics at different levels of description. In the context of religion, we could examine the reduction that takes place when one attempts to 'lift dynamics' from the scale of individual practitioners to the scale of their entire (religious) community. In this process of moving to the group-level description,

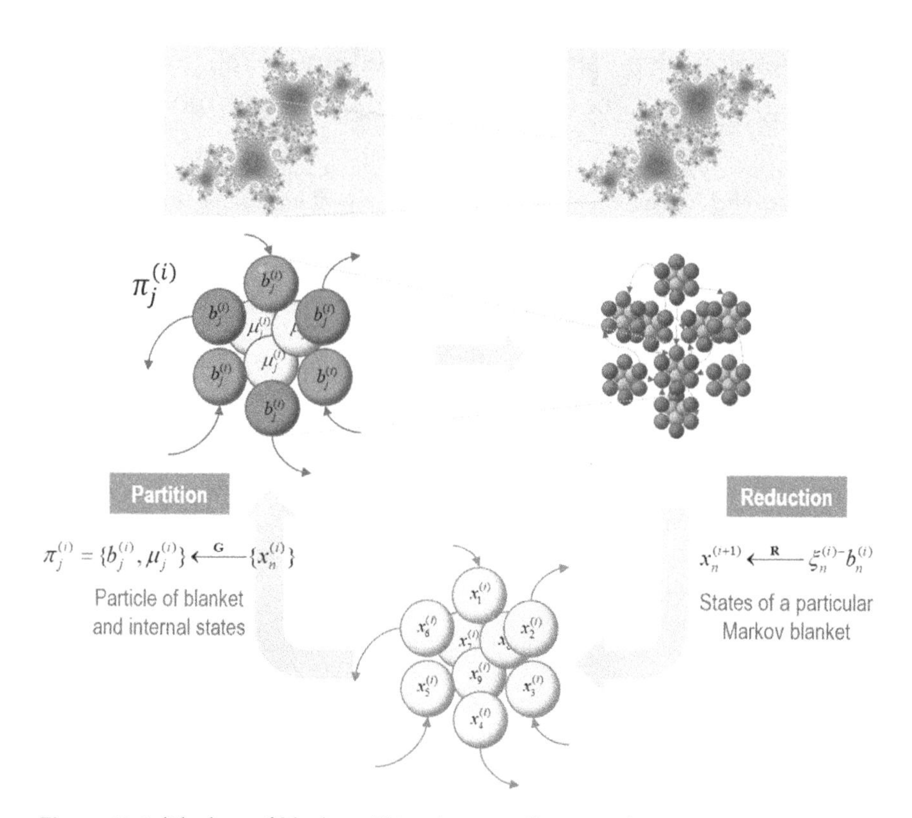

Figure 11.1 'Blankets of blankets. This schematic illustrates the recursive procedure by which successively larger (and slower) scale dynamics arise from subordinate levels. At the bottom of the figure (lower panel), we start with an ensemble of vector states (here nine). The conditional dependencies among these vector states (i.e., eigenstates) define a particular partition into particles (upper panels). Crucially, this partition equips each particle with a bipartition into blanket and internal states, where blanket states comprise active (red) and sensory states (magenta). The behaviour of each particle can now be summarised in terms of (slow) eigenmodes or mixtures of its blanket states – to produce eigenstates at the next level or scale. These constitute an ensemble of vector states and the process starts again. Formally, one can understand this in terms of coarse-graining the dynamics of a system via two operators. The first uses the particular partition to group subsets of states (**G**), while the second uses the eigenmodes of the resulting blanket states to reduce dimensionality (**R**). The upper panels illustrate the bipartition for a single particle (left panel) and an ensemble of particles; i.e., the particular partition per se (right panel). The insets on top illustrate the implicit self-similarity of particular dependencies pictorially, in moving from one scale to the next. Please see the main text of Friston (2019) for a definition of the variables used in this figure.' Reproduced with the original legend from Friston (2019, p40). © Karl Friston.

one is left with the macroscopic dynamics of their blanket states, as expressed in their religious practices. In other words, the internal states of any individual person can only influence group-level dynamics if there is (1) a consistent coupling to one's expressions (through active states) and (2) one makes sure others observe them (through sensory states). Such influences can be further amplified if (3) the interactive loop is closed with some degree of consistency that translates across individuals (i.e. partial commensurability). In complex systems modelling, such consistency takes the form of generalized synchrony[77] and can be seen as the emergence of a 'language game' in the Wittgensteinian sense.

We now elaborate on the specific mechanisms that (1) could allow individuals to self-organize into collective (semi-)coherent religious practices and (2) how this relates to the subjective experience of practitioners. We can do so by building on the framework of active inference, which has shown to be fruitful for developing models that characterize cross-scale self-organization in multi-agent systems, including morphogenesis,[78] neural packet formation[79] and ant colony behaviour.[80] In the context of sociocultural dynamics, further developments of deep active inference have been made towards the simulation of epistemic communities[81] and cumulative culture.[82]

In the last-mentioned work, Kastel & Hesp (2021) particularly aimed to incorporate the relation to subjective experiences as well, simulating 'cultural dynamics' along with the affective dimension of valence,[83] which characterizes the positive versus negative subjective value associated with one's experience. Within these simulations, that affective quality was (partially) expressed by agents along with their content-specific expressions, affording a degree of social contagion through (implicit) associative learning. In other lines of work, deep-parametric active inference has shown to be a fruitful framework for characterizing subjective experiences ranging from attentional control and meta-awareness in the context of meditative practice[84] to altered sense of self and agency.[85] As such, it has been argued that this approach provides a relatively model-agnostic methodological basis for computational phenomenology of experience.[86] As such deep active inference has been heralded as a framework with enough epistemic flexibility to provide researchers of various backgrounds with a common language in their efforts to weave together the variety of leading theories of consciousness that have emerged, integrating the even more diverse array of associated experimental findings.[87]

For the sake of clarity we will illustrate how these strands of research could be integrated to formulate a generative model that characterizes both religious practice and experience, albeit in an inevitably simplified manner, in relation to the four aspects of religion described in our introduction: self-relation, salvation, devotion and veneration. To that end, we build on the generative model presented by Ciaunica and colleagues (2021),[88] who described how higher-level states of attention and attenuation can affect one's multi-modal perception of oneself and other agents within the environment. It is relatively straightforward to expand the higher-level state space of this generative model to characterize the different ways in which its lower levels can be instantiated. Furthermore, it is possible to identify functional correspondences between this minimal setup and the different aspects of religion we discussed. The result of this exercise is shown in the form of a directed acyclic graph (DAG) in Figure 11.2. It should be noted that the highest embedding level ('Context', outermost embedding in Figure 11.2) is often interpreted as implicit, 'precognitive', and usually changing only slowly (if at all), indicative of its mostly presuppositional nature. However, it can be directly affected and even exhibit fast, non-linear transitions, because every

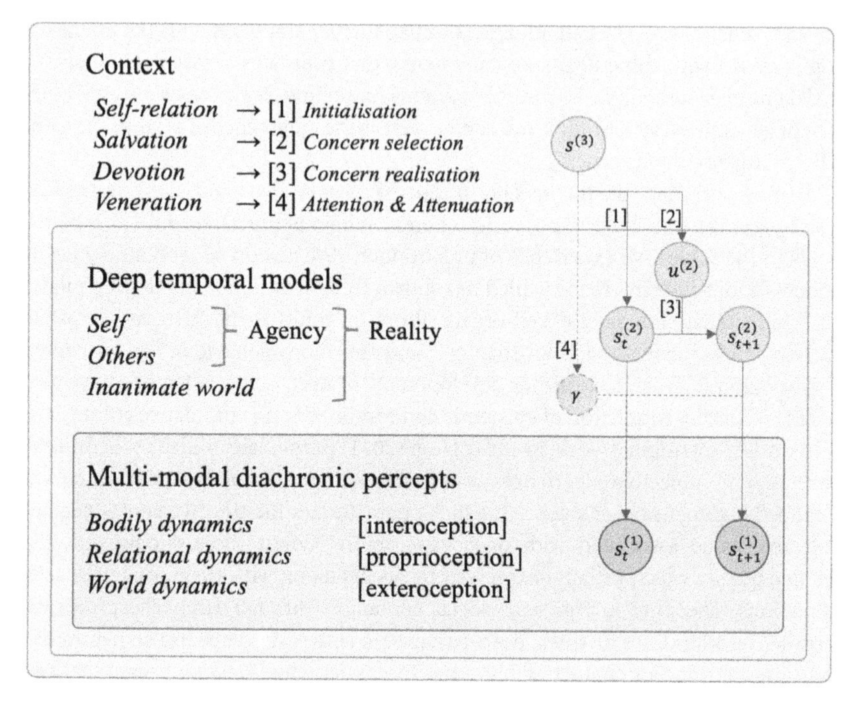

Figure 11.2 A directed acyclic graph showing a simplified heterarchical generative model of religious practice and experience, based on the canonical template of active inference, in a graphical format created based on a template from Ciaunica et al. (2022; Figure 11.1). This diagram illustrates how four important aspects of religious practices could be interpreted in the context of generative modelling. First, self-relation is taken to correspond to [1] initialization of the deep temporal generative models (indicated as middle embedding; $s^{(2)}$) of oneself and other agents within an inanimate world, that are recursively intertwined with multi-modal diachronic (i.e. temporally extended) percepts (indicated as innermost embedding; $s^{(1)}$) of bodily, relational and world dynamics (roughly corresponding to intero-, proprio and exteroception). Second, salvation is taken to correspond to [2] concern selection in the delineation of target states and admissible state transitions ($u^{(2)}$). Third, devotion can then be characterized as [3] concern realization in the enactment of state transitions that are believed to correspond to salvation for oneself and fellow practitioners. Fourth, veneration then corresponds to a form of [4] attention & attenuation, as previously formalized in terms of precision-weighting (highlighted in the diagram using dashed lines). © Casper Hesp. Based on figure from Ciaunica et al. 2022.

unidirectional arrow in this Bayesian network effectively characterizes a bidirectional relationship between the variables (in contrast to unidirectional MM). For the sake of our discussion, we will focus on explaining how it could be used to interpret each of the four aspects of religion highlighted in our introduction:

Self-relation (indicated with [1] in Figure 11.2). In a modelling context, this fundamental aspect could be interpreted as corresponding to the initial structure of the potential state spaces and their relations, which jointly characterize a generative model. This emergent notion of selfhood has been characterized as arising from dialectical attunement with one's social environment.[89]

Salvation. In a modelling context, this aspect could be interpreted as concern selection (see [2] in Figure 11.2), because it involves modulating the kinds of states that an individual would tend to be concerned with, and the kinds of solutions they might consider. In other words, concern selection would be equivalent to setting target states and acceptable ways of getting there.

Devotion. In this context, devotion could be seen as concern realization (see [3] in Figure 11.2), as it is about implementing one's intentions associated with salvation. This process tends to be community-oriented in the form of communal devotion – a crucial link between individual choices and group-level dynamics.

Veneration. This aspect could be taken to correspond to a form of attention modulation ([4] in Figure 11.2) as formalized in terms of precision-weighting, where those who are perceived as 'good' examples of devotion are praised and emulated as role model candidates that are recursively shaping the language-game, both in terms of the methods of comparison and the corresponding norms. On the flip side, those who are perceived as particularly bad examples can be punished and marginalized (or even excommunicated) as a form of contrastive learning. This distinction shows how both positive and negative valence can amplify attentional focus.

For more detailed descriptions, see Figure 11.2. While the graphical form of this model exhibits a hierarchical embedding structure, it is neither strictly hierarchical nor strictly flat because it involves cross-layer connections that manipulate the relations between layers (e.g. the dashed arrows). This kind of heterarchical structure is typical of complex adaptive systems exhibiting cross-scale metastability (or self-organized criticality) and hence resisting reduction to any particular level of description. For example, consider all the various changes that would co-occur across these levels for someone who is currently converting to a particular religion (or deconverting, for that matter). Their forms of living would tend to be affected on various levels of description, none of which can be disregarded if one aims to provide a perspicuous presentation of that (de-)conversion experience. As such, what we are presenting here is not an explanation of the phenomena of religious experiences and practices per se, but rather a toolbox that is supposed to aid researchers in the development of perspicuous presentations of this phenomenon.

11.6 On the complex relationship between religion and science: The challenging case of ethics

Previously we have argued that religious practices are 'not even wrong' from a scientific perspective – unless such religious practices involve practitioners wading into the territory of science and exposing themselves to scientific scrutiny. Throughout written history, religious leaders of various stripes have attempted to teach their fellow practitioners to avoid embedding claims about the natural world in essential religious doctrines.[90] In principle, that approach affords a peaceful coexistence of science and religion as enculturated practices, while also highlighting potential friction. For example, certain modern-day religious communities continue to embrace claims about nature that contradict scientific consensus, rejecting well-established scientific hypotheses such as, most commonly, universal common ancestry in the context of biological evolution.

While such confusion highlights the differences between science and (certain conservative strands of) religion, their language-games are both dependent on human communities. They are both rule-following and communal in nature,[91] although their attitudes towards 'goals' and 'rules' address different facets of life. Intersections have to be handled with special care because inadvertent mixing is guaranteed to lead to confusion in the Wittgensteinian sense.

The field of ethics is a useful example that highlights both the symmetries and the asymmetries inherent to the coexistence of religion and science. On the one hand, such coexistence allows people with diverse opinions about the natural world to enjoy communion in religious practices, as long as they conform to shared morals and religious doctrines. For example, doctors can participate in religious gatherings without first having to exorcise all erroneous beliefs about human body function. On the other hand, scientists with diverse religious and secular affiliations can – in principle – work together effectively on scientific research, as long as they adhere to common principles of science. However, scientific conduct tends to be additionally constrained by moral principles and ethical considerations, which themselves are outside the purview of science and tend to be influenced by religious considerations. Consider, for example, how embryonal research is constrained by the field of bioethics and moral principles such as the sanctity of life. Striking a proper balance between the tensions generated by these fronts requires yet another kind of language-game, one that has been targeted by ethicists. Even when mastered to perfection, their game would be an extremely delicate and frustrating one that requires balancing a perpetual tug-of-war between domain experts, policymakers, politicians and the general public to achieve (some degree of) convergence on decidedly non-scientific questions (e.g. concerning the relative prioritization of animal welfare in scientific experiments). The resulting answers depend on (largely incommensurable) scientific knowledge and will end up constraining scientific practices, while each of the 'players' (including the scientists) will require guidance from their individual moral principles (whether religiously motivated or not). In practice, these complex ambiguities can quickly create gridlock situations, which have been avoided by exploiting disparities in executive power, often in favour of short-term political considerations.

While a conscientious separation between the language-games of religious practices and those of the scientific method provides a partial solution that resolves many common confusions in public and private discourse, it does not provide us with a 'perspicuous presentation' of their touchpoints. It is those points of interaction that tend to give rise to significant and societally impactful ambiguities, of which the domain of ethics is perhaps the most obvious and general one. The touchpoints between science and religion can be interpreted in light of the generative model presented in Figure 11.2. Whenever possible, scientists aim to constrain themselves to develop mechanistic models (middle embedding in Figure 11.2) for what is observable, quantifiable, testable, replicable et cetera (innermost embedding in Figure 11.2).

Beyond adherence to scientific principles (outermost embedding in Figure 11.2), it does not prescribe any particular moral values or target states ([2] in Figure 11.2). However, given some variables of moral relevance and their relative importance (e.g. survival and recovery rates, self-reported quality of life), science could help inform implementation ([3] in Figure 11.2) and constrain its practices accordingly. Such considerations are particularly relevant when considering the potential benefits of experimental studies and balancing them against expected costs, an ongoing challenge for the field of ethics. This concludes our section, showing how the graphical language of generative modelling could be used to develop perspicuous presentations of the complexities intrinsic to ethical policy-making. By doing this, we aimed to demonstrate to the reader how they could take advantage of the conceptual machinery associated with generative modelling for their own use cases.

11.7 Conclusion

This chapter has taken inspiration from Wittgenstein's thinking for constructing a non-reductive toolbox associated with generative modelling, for the understanding of religion. Our treatment converged on a communal perspective on religion, suggesting it both enables and requires the emergence of implicit (or 'ineffable') meanings that are experienced both individually and collectively, with only partial commensurability. We elaborated on both the collective and perspectival aspects of religion that are incommensurable with the scientific method due to the epistemic separation between the different kinds of 'language games' of religious and scientific practices. Touchpoints between these different kinds of language-games tend to give rise to perpetual confusion and unproductive discussions. As a particularly impactful example, we discussed ethical considerations in policy-making, which demonstrates how our account can shed light on complex societal challenges.

Notes

1 Dawson and Russello 1948; Hordern 2016; 2020; Luckmann 2019; Ammerman 2020.
2 Hick 1988; Artigas 1999; Berry and Berry 2009; Hanley, Bennett and Ratcliffe 2014; Peterson and Venema 2021.
3 James 1901-1902/2003, 53, 125, 137, 443, 502; see also Turner 2008; Weed 2020; Fabb 2021; Bennett-Hunter 2022.
4 Augustine, 397 AD/1996; Kennedy, Kanthamani and Palmer 1994; Brown 2007; Burnaby 2007; Classen 2018; Appel et al. 2020; Skrzypińska 2021.
5 Smart and Murthy 1996; Pollack, Müller and Pickel 2012; Fehige 2016; Allen 2018; Barrett 2022.
6 Moore and Klein 1977.
7 See Barrett 2022.
8 Löffler 2018; Matheson 2018; McGrath 2018; Penelhum 2019; Tillson 2022.

9 Boyer 2001, 2018; Sperber and Hirschfeld 2007; Purzycki and Sosis 2010; Tremlin 2010.
10 Wood and Shaver 2018, 1.
11 Ryle 1949; Hutto and Myin 2017; Di Paolo, Cuffari and De Jaegher 2018; Gallagher 2020.
12 Ryle 1949; Hutto and Myin 2017; Di Paolo, Cuffari and De Jaegher 2018; Gallagher 2020.
13 Fodor 1983.
14 Boyer 2001, 144.
15 Boyer 2001, 17.
16 For example Plantinga 1967.
17 See for example Bennett and Hacker 2022, 179.
18 Driver and Spence 2000; Baltieri and Buckley 2018; Khambhati et al. 2018; Zerilli 2019; Pietraszewski and Wertz 2022.
19 Asprem 2019; Schjødt and van Elk 2019; Szocik and Van Eyghen 2021.
20 Asprem 2019.
21 Kalén et al. 2021, Nair et al. 2021; Mononen, Niemivirta and Korhonen, 2022; Niemivirta and Korhonen 2022; Saito et al. 2022.
22 Rao and Ballard 1999; Spratling 2016; Smith, Badcock and Friston 2021; Kramer et al. 2022; Pezzulo, Parr and Friston 2022.
23 Anderson 2019; Asprem 2019; see other work pointing in the same direction Van Elk and Wagenmakers 2017; Schjødt 2018; Schjød and van Elk 2019; Szocik and Van Eyghen 2021.
24 Wittgenstein's works are cited by their standard abbreviations. PI = Philosophical Investigations, OC = On Certainty, RFGB = Remarks on Frazer's Golden Bough, BB = Blue Book, LE = Lecture on Ethics; LRB = Lectures on Religious Belief.
25 RFGB, 119.
26 RFGB, 119.
27 Woo 2014.
28 Wittgenstein, LRB, 55.
29 BB, 18.
30 James 1901–02/2003; Turner 2008; Weed 2020; Fabb 2021; Bennett-Hunter 2022.
31 RFGB, 121.
32 RFGB, 122–3.
33 LE, 43.
34 RFGB, 119.
35 See *PI*, §109.
36 *PI*, §109.
37 *PI*, §114.
38 *OC*, §337.
39 *OC*, §204.
40 *RPP* I, 630.
41 *PI*, §142.
42 *RFM*, §92.
43 *OC*, §358.
44 *OC*, §359.
45 Moyal-Sharrock 2015, 2016.
46 For example Gettier 1963.
47 Stanley and Williamson 2017.

48 Pavese 2019.
49 In other work (Hipólito et al. 2021), we have disputed this.
50 *PI*, §201.
51 Gettier 1963.
52 *BB*, 34.
53 *PI*, §198.
54 *OC*, §192.
55 *OC*, §204.
56 *PI*, §202.
57 Moyal-Sharrock 2021.
58 Moyal-Sharrock 2017.
59 Richardson and Uebel 2007.
60 Grimmel 2015.
61 Grimmel 2015.
62 Grimmel 2015, 6.
63 Rhees 1965; Burbules, Smeyers 2003; Bizri 2021.
64 Altmann 1987; Philips 1993; Diamond 2019; Asad 2020.
65 Shiner 1974; Perloff 1994; Schroeder 2020.
66 Birgani and Soqandi 2020.
67 *PI*, §7.
68 Arrington 2001.
69 *Zettel*, §717.
70 Note that as stressed previously in this chapter, not all religious claims are nonsensical. It is also important to note that, as it is widely known, Wittgenstein's thinking has evolved from the *Tractatus*, putting emphasis on *nonsensical*, towards *Philosophical Investigations*, elucidating upon *grammatical* statements that can be made about diverse forms of experiencing/engaging with the world, for example aesthetic, ethical, religious, psychological.
71 Lurie 2012.
72 Arrington 2001, 172.
73 'we must plow through the whole of language', RFGB, 131.
74 'Complex Systems Theory (CST), as a formalism, is widely known for rejecting *the machine vocabulary of processing devices, programs, storage units, schemata, modules.* . . . We substitute, instead, a vocabulary suited to fluid, organic systems, with certain thermodynamic properties'(Thelen and Smith 1996, xix, emphasis added). Complex systems are systems composed of many interacting elements, which makes them extremely rich dynamical systems showing a huge range of phenomena with properties that arise from these relationships, such as nonlinearity, emergence, spontaneous order, adaptation and feedback loops, amongst others.
75 See for example Hesp et al. 2019; Hipólito et al. 2021.
76 2019, 5.
77 See, for example. Friston and Frith 2015; Kastel and Hesp 2021; Hipólito and van Es 2022.
78 For example Friston et al. 2015; Kuchling et al. 2020.
79 For example Ramstead et al. 2021.
80 Friedman et al. 2021.
81 Albarracin et al. 2022.
82 Kastel and Hesp 2021.
83 As derived by Hesp et al. 2021.

84 Sandved-Smith et al. 2021.
85 Ciaunica, Safron and Delafield-Butt 2021.
86 Ramstead et al. 2021.
87 See for example Barret 2017; Saffron 2020; Kirchhoff and Kiverstein 2020; Friston, Wiese, and Hobson 2020; Deane 2021; Ciaunica, Saffron, and Delafield-Butt 2021; Vilas, Auksztulewicz, and Melloni 2021; Whyte, Howhy, and Smith 2022; Seth and Bayne 2022; Seth 2021.
88 See, in particular, their Figure 1.
89 See for example Bolis and Schilbach 2020.
90 See, for example, Augustine in *Confessions*, Chapter 5, Section 5, 397 AD.
91 Harré 2001; Bell 1969.

References

Adams, A. (1982), 'Review of W. M. Indich (1982). *Consciousness in Advaita Vedanta*'. *Philosophy East and West*, 32 (4): 468–70.

Addis, M. (2013), 'Wittgenstein and the Naturalness of Religious Belief', in A. Geertz (ed.), *Origins of Religion, Cognition and Culture*, 258–68, Durham: Acumen Publishing.

Airini, M. Anae, K. Mila-Schaaf, E. Coxon, D. Mara and K. Sanga (2010), *TEU LE VA–Relationships Across Research and Policy in Pasifika Education: A Collective Approach to Knowledge Generation & Policy Development for Action Towards Pasifika Education Success*, edited by R. Carolissen, P. Duckett, C. Sonn and N. E. Allen, 2nd edn, vol. 62, New Zealand: Ministry of Education.

Albarracin, M., D. Demekas, M. J. Ramstead and C. Heins (2022), 'Epistemic Communities under Active Inference', *Entropy*, 24 (4): 476.

Albinus, L., J. G. F. Rothhaupt and A. Seery, eds. (2016), *Wittgenstein's Remarks on Frazer: The Text and the Matter*, Berlin: De Gruyter.

Ali, F. and T. Chamorro-Premuzic (2010), 'Investigating Theory of Mind Deficits in Nonclinical Psychopathy and Machiavellianism', *Personality and Individual Differences*, 49 (3): 169–74.

Allen, D. B., ed. (2018), *Culture and Self: Philosophical and Religious Perspectives, East and West*, New York: Routledge.

Altmann, A. (1987), 'The God of Religion, the God of Metaphysics and Wittgenstein's "Language Games"', *Zeitschrift für Religions-und Geistesgeschichte*, 39 (4): 289–306.

Ambrose, A., ed. (1979), *Wittgenstein's Lectures, Cambridge 1932–35*, Totowa: Littlefield and Adams (AWL).

Ammerman, N. T. (2020), 'Rethinking Religion: Toward a Practice Approach', *American Journal of Sociology*, 126 (1): 6–51.

Andersen, M. (2017), 'Predictive Coding in Agency Detection', *Religion, Brain and Behavior*, 9 (1): 1–20.

Andersen, M. (2019), 'Predictive Coding in Agency Detection', *Religion, Brain & Behavior*, 9 (1): 65–84.

Andrejč, G. (2015), 'Reading Wittgenstein (on Belief) with Tillich (on Doubt)', *Neue Zeitschrift für Systematische Theologie und Religionsphilosophie*, 57 (1): 60–86.

Andrejč, G. (2016), *Wittgenstein and Interreligious Disagreement: A Philosophical and Theological Perspective*, Basingstoke: Palgrave-Macmillan.

Andrejč, G. and D. Weiss (2019), *Interpreting Interreligious Relations with Wittgenstein*, Leiden: Brill.

Appel, J. E., C. L. Park, J. H. Wortmann and H. T. van Schie (2020), 'Meaning Violations, Religious/Spiritual Struggles, and Meaning in Life in the Face of Stressful Life Events', *The International Journal for the Psychology of Religion*, 30 (1): 1–17.

Aristotle (1956), *De Anima / On the Soul* (trans. W. D. Ross), in Jonathan Barnes, ed. (1984), *The Complete Works of Aristotle*, Princeton: Princeton University Press.

Arno, A. (1976), 'Ritual of Reconciliation and Village Conflict Management in Fiji', *Oceania*, 47 (1): 49–65.

Arrington, R. (2001), '"Theology as grammar": Wittgenstein and Some Critics', in M. Addis and R. L. Arrington (eds), *Wittgenstein and Philosophy of Religion*, 167–84, London: Routledge.
Arrington, R. and M. Addis, eds. (2001), *Wittgenstein and Philosophy of Religion*, London: Routledge.
Artigas, M. (1999), *The Mind of the Universe: Understanding Science and Religion*, Radnor PA: Templeton Foundation Press.
Asad, T. (2003), *Formations of the Secular: Christianity, Islam, Modernity*, Redwood City CA: Stanford University Press.
Asad, T. (2020), 'Thinking About Religion Through Wittgenstein', *Critical Times*, 3 (3): 403–42.
Asprem, E. (2019), 'Predictive Processing and the Problem of (Massive) Modularity', *Religion, Brain and Behavior*, 9 (1): 84–6.
Atran, S. (2002), *In Gods We Trust: The Evolutionary Landscape of Religion*, Oxford: Oxford University Press.
Atran, S. (2006), 'The Scientific Landscape of Religion: Evolution, Culture, and Cognition', in Philip Clayton and Zachory Simpson (eds), *The Oxford Handbook of Religion and Science*, 407–29, Oxford: Oxford University Press.
Atran, S. and A. Norenzayan (2004), 'Religion's Evolutionary Landscape: Counterintuition, Commitment, Compassion, and Communion', *Behavioral and Brain Sciences*, 27: 713–70.
Atran, S., A. Norenzayan, J. Falkner and M. Schaller (2006), 'Memory and Mystery: The Cultural Selection of Minimally Counterintuitive Narratives', *Cognitive Science*, 30 (3): 531–53.
Augustine, S. (1996), *De Doctrina Christiana*, Oxford: Clarendon Press.
Axtell, G. (2019), *Problems of Religious Luck: Assessing the Limits of Reasonable Religious Disagreement*, Lanham: Lexington Books.
Axtell, G. (2020a), 'Review of Gorazd Andrejč (2016) *Wittgenstein and Inter-religious Disagreement*', *Wittgenstein-Studien*, 11 (1): 285–90.
Axtell, G. (2020b), 'An Argument for the Conceptual Incoherence of Mutualist Exclusivism', *Academia Letters*: Article 94.
Axtell, G. (2022a), 'Wittgenstein and the ABC's of Religious Epistemics' [forthcoming].
Axtell, G. (2022b), 'Constructive Engagement and Its Discontents: Distinguishing Comparative Philosophy from Post-liberal Analytical Apologetics', in Martin Ovens (ed.), *What Is Comparative Philosophy?*, Oxford: Oxford Scholars.
Axtell, G. (2022c), 'Theological Walls, Insularity, and the Prospects for Global Philosophy', *12th East-West Philosophers Conference* presentation draft paper, under review.
Azari, N. P., J. Nickel, G. Wunderlich, M. Niedeggen, H. Hefter, L. Tellmann, H. Herzog, P. Stoerig, D. Birnbacher and R. J. Seitz (2001), 'Neural Correlates of Religious Experience', *European Journal of Clinical Pharmacology*, 13 (8): 1649–52.
Baghramian, M. and A. Coliva (2020), *Relativism*, London: Routledge.
Baker, G. (2004), 'Wittgenstein's Method and Psychoanalysis', in K. Morris (ed.), *Wittgenstein's Method: Neglected Aspects*, 205–22, Oxford: Wiley Blackwell.
Baker, G. P. and P. M. S. Hacker (2005a), 'Philosophy', in *Wittgenstein: Understanding and Meaning, vol. 1 of an Analytical Commentary on the Philosophical Investigations, Part I: Essays*, 2nd edn, 271–306, Oxford: Wiley Blackwell.
Baker, G. P. and P. M. S. Hacker (2005b), *Wittgenstein: Understanding and Meaning. Part II: Exegesis*, Second, ext. rev. edn, Malden: Blackwell Publishing.
Baker, G. P. and P. M. S. Hacker (2009), *Wittgenstein: Rules, Grammar and Necessity. Essays and Exegesis of 185–242*, Second, ext. rev. edn, Malden: Wiley-Blackwell.

Baltieri, M. and C. L. Buckley (2018), 'The Modularity of Action and Perception Revisited using Control Theory and Active Inference'. arXiv Preprint ArXiv:1806.02649.
Bangu, S. (2018), *Naturalizing Logico-Mathematical Knowledge: Perspectives from Philosophy, Psychology, and Cognitive Science*, New York: Routledge.
Banreti-Fuchs, K. M. (1967), 'Perception Without Awareness', *Acta Psychologica*, 26 (2): 148–60.
Barbour, I. (1990), *Religion in an Age of Science*, Glasgow: HarperCollins.
Bareham, C. A., M. Oxner, T. Gastrell and D. Carmel (2020), 'Beyond the Neural Correlates of Consciousness: Using Brain Stimulation to Elucidate Causal Mechanisms Underlying Conscious States and Contents', *Journal of the Royal Society of New Zealand*, 51 (1): 143–70.
Baron-Cohen, S. (1995), *Mindblindness: An Essay on Autism and Theory of Mind*, Cambridge MA: MIT Press.
Baron-Cohen, S. and S. Wheelwright (2004), 'The Empathy Quotient: An Investigation of Adults with Asperger Syndrome or High Functioning Autism, and Normal Sex Differences', *Journal of Autism and Developmental Disorders*, 34 (2): 163–75.
Barrett, H. C., A. Bolyanatz, A. N. Crittenden, D. M. T. Fessler, S. Fitzpatrick, M. Gurven, J. Henrich, M. Kanovsky, G. Kushnick, A. Pisor, B. Scelza, S. Stich, C. von Rueden, W. Zhao and S. Laurence (2016), 'Small-Scale Societies Exhibit Fundamental Variation in the Role of Intentions in Moral Judgment', *PNAS*, 113 (17): 4688–93.
Barrett, H. C., T. Broesch, R. M. Scott, Z. He, R. Baillargeon, D. Wu, M. Bolz, J. Henrich, P. Setoh, J. Wang and S. Laurence (2013), 'Early False-Belief Understanding in Traditional Non-Western Societies', *Proceedings of the Royal Society of London. Series B*, 280 (1755): 1–6.
Barrett, J. L. (2000), 'Exploring the Natural Foundations of Religion', *Trends in Cognitive Science*, 4 (1): 29–34.
Barrett, J. L. (2004), *Why Would Anyone Believe in God*, Plymouth: Altamira Press.
Barrett, J. L. (2011), 'Cognitive Science of Religion: Looking Back, Looking Forward', *Journal for the Scientific Study of Religion*, 50 (2): 229–39.
Barrett, J. L., ed. (2022), *The Oxford Handbook of the Cognitive Science of Religion*, Oxford: Oxford University Press.
Barrett, J. L. and I. M. Church (2013), 'Should CSR Give Atheists Epistemic Assurance? On Beer-Goggles, BFFs, and Skepticism Regarding Religious Beliefs', *The Monist*, 96 (3): 311–24.
Barrett, J. L. and F. C. Keil (1996), 'Conceptualizing a Nonnatural Entity: Anthropomorphism in God Concepts', *Cognitive Psychology*, 31 (3): 219–47.
Barrett, L. F. (2017a), *How Emotions Are Made: The Secret Life of the Brain*, Boston: Houghton Mifflin Harcourt.
Barrett, L. F. (2017b), 'The Theory of Constructed Emotion: An Active Inference Account of Interoception and Categorization', *Social Cognitive and Affective Neuroscience*, 12 (1): 1–23.
Batson, C. D. (1997), 'Self–Other Merging and the Empathy–Altruism Hypothesis: Reply to Neuberg et al. (1997)', *Journal of Personality and Social Psychology*, 73 (3): 517–22.
Beale, J. (2017), 'Wittgenstein's Anti-scientistic Worldview', in J. Beale and I. J. Kidd (eds), *Wittgenstein and Scientism*, 59–80, London: Routledge.
Beale, J. and I. J. Kidd (2017), 'Introduction: Wittgenstein and Scientism', in Jonathan Beale and Ian James Kidd (eds), *Wittgenstein and Scientism*, 1–6, New York: Routledge.
Becker, M., V. L. Vignoles, E. Owe, M. J. Easterbrook, R. Brown, P. B. Smith, S. Abuhamdeh, B. Cendales Ayala, R. B. Garðarsdóttir, A. Torres, L. Camino, M. H. Bond, G. Nizharadze, B. Amponsah, I. Schweiger Gallo, P. Prieto Gil, R. Lorente

Clemares, G. Campara, A. Espinosa, M. Yuki, X. Zhang, J. Zhang, M. Zinkeng, J. A. Villamar, E. Kusdil, S. Çağlar, C. Regalia, C. Manzi, M. Brambilla, D. Bourguignon, B. Möller, M. Fülöp, M. E. J. Macapagal, T. Pyszczynski, P. Chobthamkit, N. Gausel, P. Kesebir, G. Herman, M. Courtois, C. Harb, B. Jalal, A. Tatarko, S. Aldhafri, R. Kreuzbauer, S. H. Koller, K. H. Mekonnen, R. Fischer, T. L. Milfont, S. E. Des Rosiers, J. L. Jaafar, M. Martin, P. Baguma, S. Lv, S. J. Schwartz, A. Gavreliuc, I. Fritsche, R. González, N. Didier, D. Carrasco and S. Lay (2017), 'Being Oneself Through Time: Bases of Self-Continuity Across 55 Cultures*', *Self and Identity*, 17 (3): 276–93.

Bell, R. H. (1969), 'Wittgenstein and Descriptive Theology', *Religious Studies*, 5 (1): 1–18.

Bennett, M. R. and P. M. S. Hacker (2022), *Philosophical Foundations of Neuroscience*, 2nd edn, Hoboken: Wiley Blackwell [1st ed., 2003].

Bennett-Hunter, G. (2022), 'Mysticism, Ritual, and the Meaning of Life', *The Oxford Handbook of Meaning in Life*, 243–58, Oxford: Oxford University Press.

Benton, Matthew A. (2006), 'The Modal Gap: The Objective Problem of Lessing's Ditch(es) and Kierkegaard's Subjective Reply', *Religious Studies*, 42 (1): 27–44.

Bering, J. (2002), 'The Existential Theory of Mind', *Review of General Psychology*, 6 (1): 3–24.

Bering, J. M. (2006), 'The Folk Psychology of Souls', *Behavioral and Brain Sciences*, 29 (5): 453–98.

Bering, J. M. (2011), *The Belief Instinct: The Psychology of Souls, Destiny and the Meaning of Life*, New York: W. W. Norton and Company.

Berkes, F., J. Colding and C. Folke (2000), 'Rediscovery of Traditional Ecological Knowledge as Adaptive Management', *Ecological Applications*, 10 (5): 1251–62.

Berry, J. W. (2015), 'Global Psychology: Implications for Cross-Cultural Research and Management', *Cross Cultural Management: An International Journal*, 22 (3): 342–55.

Berry, T. and T. M. Berry (2009), *The Sacred Universe: Earth, Spirituality, and Religion in the Twenty-First Century*, New York, NY: Columbia University Press.

Biabanaki, S. M. (2020), 'The Cognitive Biases of Human Mind in Accepting and Transmitting Religious and Theological Beliefs: An Analysis Based on the Cognitive Science of Religion', *HTS Teologiese Studies/ Theological Studies*, 76 (1).

Bilimoria, P. (1980), 'Perception (Pratyaksa) in Advaita Vedanta', *Philosophy East and West*, 30 (1): 35–44.

Billings, A. J., L. R. Churchill and R. Payne (2010), 'Severe Brain Injury and the Subjective Life', *Hastings Center Report*, 40 (3): 17–21.

Birch, J. and C. Heyes (2021), 'The Cultural Evolution of Cultural Evolution', *Philosophical Transactions of the Royal Society B: Biological Sciences*, 376: 20200051, 1–8.

Bird-David, N. (1999), '"Animism" Revisited: Personhood, Environment, and Relational Epistemology', *Current Anthropology*, 40 (Supplement): S567–S591: 67–91.

Birgani, S. Z. and M. Soqandi (2020), 'Wittgenstein's Concept of Language Games', *Britain International of Linguistics Arts and Education (BIoLAE) Journal*, 2 (2): 641–7.

Bishop, J. (2007), *Believing by Faith: An Essay in the Epistemology and Ethics of Religious Belief*, New York: Clarendon Press.

Bizri, R. (2021), *Ethics After Wittgenstein: Contemplation and Critique*, edited by Richard Amesbury and Hartmut von Sass, London: Bloomsbury Academic.

Blair, R. (2008), 'Fine Cuts of Empathy and the Amygdala: Dissociable Deficits in Psychopathy and Autism', *The Quarterly Journal of Experimental Psychology*, 61 (1): 157–70.

Bloom, P. (2004), *Descartes' Baby: How the Science of Child Development Explains What Makes Us Human*, New York: Basic Books.
Bloom, P. (2007), 'Religion Is Natural', *Developmental Science*, 10: 147–51.
Bodewitz, H. W. (2019), 'The Hindu Doctrine of Transmigration: Its Origin and Background', in D. Heilijgers, J. Houben and K. van Kooij (eds), *Vedic Cosmology and Ethics*, 3–19, Leiden: Brill.
Boghossian, P. (2006), *Fear of Knowledge: Against Relativism and Constructivism*, Oxford: Oxford University Press.
Bolis, D. and L. Schilbach (2020), '"I Interact Therefore I Am": The Self as a Historical Product of Dialectical Attunement', *Topoi*, 39 (3): 521–34.
Bouwsma, O. K. (1987), *Wittgenstein: Conversations 1949–1951*, Indianapolis: Hackett.
Boyer, P. (1994), *The Naturalness of Religious Ideas*, Berkeley: University of California Press.
Boyer, P. (2001), *Religion Explained: The Evolutionary Origins of Religious Thought*, New York: Basic Books (London: Vintage Edition, 2002).
Boyer, P. (2003), 'Religious Thought and Behaviour as By-Products of Brain Function', *Trends in Cognitive Sciences*, 7 (3): 119–25.
Boyer, P. (2008), 'Religion: Bound to Believe?', *Nature*, 455 (7216): 1038–9.
Boyer, P. and P. Liénard (2006), 'Why Ritualized Behavior? Precaution Systems and Action Parsing in Developmental, Pathological and Cultural Rituals', *The Behavioral and Brain Sciences*, 29 (6): 595–613.
Boyer, P. (2018), *Minds Make Societies: How Cognition Explains the World Humans Create*, New Haven CT: Yale University Press.
Boyer, P. and P. Liénard (2020), 'Ingredients of "rituals" and Their Cognitive Underpinnings', *Philosophical Transactions of the Royal Society of London*, 375 (1805): 1–6.
Boyer, P. and C. Ramble (2001), 'Cognitive Templates for Religious Concepts: Cross-Cultural Evidence for Recall of Counter-Intuitive Representations', *Cognitive Science*, 25 (4): 535–64.
Bremmer, J. (1994), *Greek Religion*, Oxford: Oxford University Press.
Brison, K. J. (2007), *Our Wealth Is Loving Each Other: Self and Society in Fiji*, Lanham: Lexington Books.
Brown, P. R. L. (2007), *Religion and Society in the Age of St. Augustine*, Eugene, OR: Wipf and Stock Publishers.
Brune, M. (2006), 'The Evolutionary Psychology of Obsessive-Compulsive Disorder: The Role of Cognitive Metarepresentation', *Perspectives in Biology and Medicine*, 49 (3): 317–29.
Bruner, J. (1990), *Acts of Meaning*, Cambridge, MA: Harvard University Press.
Bryant, E. F. (2015), *The Yoga Sutras of Patanjali: A New Edition, Translation, and Commentary*, New York: North Point Press.
Burbules, N. C. and P. Smeyers (2003), 'Wittgenstein, the Practice of Ethics, and Moral Education', in Scott Fletcher (ed.), *Philosophy of Education 2002*, 248–57, Urbana IL: Philosophy of Education Society.
Burgess, N. (2006), 'Spatial Memory: How Egocentric and Allocentric Combine', *Trends in Cognitive Science*, 10 (12): 551–7.
Burkert, W. (1985), *Greek Religion*, Cambridge, MA: Harvard University Press.
Burley, M., ed. (2018), *Wittgenstein, Religion and Ethics*, London: Bloomsbury Publishing.
Burley, M. (2020), *A Radical Pluralist Philosophy of Religion: Cross-Cultural, Multireligious, Interdisciplinary*, New York: Bloomsbury Publishing.

Burnaby, J. (2007), *Amor Dei: A Study of the Religion of St. Augustine*, Eugene, OR: Wipf and Stock Publishers.
Buss, D. M. (1994), *The Evolution of Desire: Strategies of Human Mating*, New York: Basic Books.
Byrne, A. (2004), 'Inverted Qualia', *Stanford Encyclopedia of Philosophy* (Spring 2013 Edition). https://plato.stanford.edu/entries/qualia-inverted/.
Caldwell-Harris, C. L., A. L. Wilson, E. LoTempio and B. Beit-Hallahmi (2011), 'Exploring the Atheist Personality: Well-Being, Awe, and Magical Thinking in Atheists, Buddhists, and Christians', *Mental Health, Religion and Culture*, 14 (7): 659–72.
Callaghan, T., H. Moll, H. Rakoczy, F. Warneken, U. Liszkowski, T. Behne, M. Tomasello and W. A. Collins (2011), 'Early Social Cognition in Three Cultural Contexts', *Monographs of the Society for Research in Child Development*, 76 (2): i–v-vii-1–142.
Callaghan, T., P. Rochat, A. Lillard, M. L. Claux, H. Odden, S. Itakura, S. Tapanya and S. Singh (2005), 'Synchrony in the Onset of Mental-State Reasoning: Evidence from Five Cultures', *Psychological Science*, 16 (5): 378–84.
Callicott, C. (2013), 'Interspecies Communication in the Western Amazon: Music as a Form of Conversation Between Plants and People', *European Journal of Ecopsychology*, 4: 32–43.
Campbell, M. and M. O'Sullivan (2015), 'Wittgenstein on Perception: An Overview', in M. Campbell and M. O'Sullivan (eds), *Wittgenstein and Perception*, 10–32, Abingdon: Routledge.
Canfield, J. (2007), *Becoming Human: The Development of Language, Self, and Self-Consciousness*, New York: Palgrave Macmillan.
Carnap, R. (1963), 'Intellectual Autobiography', in P. A. Schilpp (ed.), *The Philosophy of Rudolf Carnap*, 3–43, LaSalle: Open Court.
Carnap, R. (2021), 'The Elimination of Metaphysics Through Logical Analysis of Language', in Arthur Pap (ed.), *Logical Empiricism at Its Peak*, 10–31, London: Routledge.
Carroll, T. D. (2010), 'The Debate over "Wittgensteinian Fideism" and Phillips' Contemplative Philosophy of Religion', in Ingolf U. Dalferth and Hartmut von Sass (eds), *The Contemplative Spirit: D. Z. Phillips on Religion and the Limits of Philosophy*, 99–114, Zürich: Mohr Siebeck.
Carroll, T. D. (2014), *Wittgenstein Within the Philosophy of Religion*, Basingstoke: Palgrave-Macmillan.
Carroll, T. D. (2017), 'Clarifying Conversations: Understanding Cultural Difference in Philosophical Education', in M. A. Peters and J. Stickney (eds), *A Companion to Wittgenstein on Education: Pedagogical Investigations*, 757–69, Singapore: Springer.
Carroll, T. D. (2019), 'Wittgenstein and Ascriptions of "Religion"', in Gorazd Andrejč and Daniel H. Weiss (eds), *Interpreting Interreligious Relations with Wittgenstein: Philosophy, Theology and Religious Studies*, 54–72, Leiden: Brill.
Carsten, J. (1995), 'The Substance of Kinship and the Heat of the Hearth: Feeding, Personhood, and Relatedness Among Malays in Pulau Langkawi', *American Ethnologist*, 22 (2): 223–41.
Castro, E. V. de (1998), 'Cosmological Deixis and Amerindian Perspectivism', *The Journal of the Royal Anthropological Institute*, 4 (3): 469–88.
Castro, E. V. de (2004), 'Exchanging Perspectives: The Transformation of Objects into Subjects in Amerindian Ontologies', *Common Knowledge*, 10 (3): 463–84.
Cavell, S. (1978), 'What Becomes of Things on Film?', *Philosophy and Literature*, 2 (2): 249–57.

Chakroff, A. and L. Young (2015), 'How the Mind Matters for Morality', *AJOB Neuroscience*, 6 (3): 43–8.
Chaves, M. (2010), 'SSSR Presidential Address: Rain Dances in the Dry Season: Overcoming the Religious Congruence Fallacy', *Journal for the Scientific Study of Religion*, 49 (1): 1–14.
Cheney, D. and R. Seyfarth (2007), *Baboon Metaphysics: The Evolution of Social Mind*, Chicago: University of Chicago Press.
Chillingworth, W. (1638), *Religion of Protestants*, Oxford: L. Lichfield.
Choe, K. S., F. C. Keil and P. Bloom (2012), 'Developing Intuitions About How Personal and Social Properties Are Linked to the Brain and the Body', *Infant and Child Development*, 21 (4): 430–41.
Choo, F. (2018), 'The Epistemic Significance of Religious Disagreements: Cases of Unconfirmed Superiority Disagreements', *Topoi*, 40 (5): 1139–47.
Chryssides, G. D. and S. E. Gregg, eds. (2019), *The Insider/Outsider Debate: New Perspectives in the Study of Religion*, Sheffield: Equinox Publishing.
Chudek, M., R. A. McNamara, S. Birch, P. Bloom and J. Henrich (2017), 'Do Minds Switch Bodies? Dualist Interpretations Across Ages and Societies', *Religion, Brain and Behavior*, 83 (2): 1–15.
Cialdini, R. B., S. L. Brown, B. P. Lewis, C. L. Luce and S. L. Neuberg (1997), 'Reinterpreting the Empathy-Altruism Relationship: When One into One Equals Oneness', *Journal of Personality and Social Psychology*, 73 (3): 481–94.
Ciaunica, A., A. Safron and J. Delafield-Butt (2021), 'Back to Square One: The Bodily Roots of Conscious Experiences in Early Life', *Neuroscience of Consciousness*, 7 (2): 1–10.
Ciaunica, A., A. Seth, J. Limanowski, C. Hesp and K. J. Friston (2022), 'I Overthink—Therefore I Am Not: An Active Inference Account of Altered Sense of Self and Agency in Depersonalisation Disorder', *Consciousness and Cognition*, 101: 1–14.
Clack, B. R. (1999), *Wittgenstein, Frazer, and Religion*, London: Palgrave-Macmillan.
Clack, B. R. (2001), 'Wittgenstein and Magic', in R. L. Arrington and M. Addis (eds), *Wittgenstein and Philosophy of Religion*, 12–28, London: Routledge.
Clark, A. (1997), *Being There: Putting Brain, Body, and World Together Again*, Cambridge, MA: MIT Press.
Clark, A. (2013), 'Whatever Next? Predictive Brains, Situated Agents, and the Future of Cognitive Science', *Behavioral and Brain Sciences*, 36 (3): 181–204.
Classen, A. (2018), 'The Human Quest for Happiness and Meaning: Old and New Perspectives: Religious, Philosophical, and Literary Reflections from the Past as a Platform for Our Future: St. Augustine, Boethius, and Gautier de Coincy', *Athens Journal of Humanities & Arts*, 5 (2): 179–206.
Cohen, A. B. and P. C. Hill (2007), 'Religion as Culture: Religious Individualism and Collectivism Among American Catholics, Jews, and Protestants', *Journal of Personality*, 75 (4): 709–42.
Coliva, A. (2017), 'Rituals, Philosophy, Science, and Progress: Wittgenstein on Frazer', in Jonathan Beale and Ian James Kidd (eds), *Wittgenstein and Scientism*, 39–58, New York: Routledge.
Collier, M. (2011), 'Hume's Science of Emotions: Feeling Theory Without Tears', *Hume Studies*, 37 (1): 3–18.
Collier, M. (2014), 'The Natural Foundations of Religion', *Philosophical Psychology*, 27 (5): 665–80.
Cook, G., ed. (1997), *Crosscurrents in Indigenous Spirituality: Interface of Maya, Catholic and Protestant Worldviews*, Leiden: Brill.

Craig, W. L. and W. Harris (2008), 'Lessing's Broad Ugly Ditch: A Conversation with William Lane Craig', *Reasonable Faith* Blog podcast interview. Available online: https://www.reasonablefaith.org/media/reasonable-faith-podcast/lessings-broad-ugly-ditch (Accessed March 2020).

Crane, T. (2017), *The Meaning of Belief: Religion from an Atheist's Point of View*, Cambridge, MA: Harvard University Press.

Crane, T. and C. French (2016), 'The Problem of Perception', *Stanford Encyclopedia of Philosophy*. Available online: https://plato.stanford.edu/entries/perception-problem/.

Crary, A. (2005), 'Wittgenstein and Ethics: A Discussion with Reference to on Certainty', in D. Moyal-Sharrock and W. H. Brenner (eds), *Readings of Wittgenstein's on Certainty*, 275–301, London: Palgrave Macmillan.

Crary, A. and R. Read, eds. (2000), *The New Wittgenstein*, London: Routledge.

Crick, F. (1995), *The Astonishing Hypothesis*, London: Touchstone.

Curtin, C. M., H. C. Barrett, A. Bolyanatz, A. N. Crittenden, D. M. T. Fessler, S. Fitzpatrick, M. Gurven, M. Kanovsky, G. Kushnick, S. Laurence, A. Pisor, B. Scelza, S. Stich, C. von Rueden and J. Henrich (2020), 'Kinship Intensity and the Use of Mental States in Moral Judgment Across Societies', *Evolution and Human Behavior*, 41 (5): 415–29.

D'Andrade, R. (1987), 'A Folk Model of the Mind', in D. Holland and N. Quinn (eds), *Cultural Models in Language and Thought*, 112–48, Cambridge: Cambridge University Press.

Damasio, A. R. (1994), *Descartes' Error. Emotion, Reason, and the Human Brain*, New York: Penguin Books.

Danziger, K. (1997), *Naming the Mind: How Psychology Found Its Language*, Thousand Oaks: Sage Publications.

Das, V. (2020), *Textures of the Ordinary: Doing Anthropology After Wittgenstein*, New York: Fordham University Press.

Davis, M. H. (1980), 'A Multidimensional Approach to Individual Differences in Empathy', *JSAS Catalog of Selected Documents in Psychology*, 10: 85–104.

Davis, M. H. (1983), 'Measuring Individual Differences in Empathy: Evidence for a Multidimensional Approach', *Journal of Personality and Social Psychology*, 44 (1): 113–26.

Dawkins, R. (2016 [1976]), *The Selfish Gene: 40th Anniversary Edition*, Oxford: Oxford University Press.

Dawson, C. and G. J. Russello (1948), *Religion and Culture (13)*, London: Sheed & Ward.

Day, M. (2004), 'Religion, Off-line Cognition and the Extended Mind', *Journal of Cognition and Culture*, 4 (1): 101–21.

de Ridder, J. (2018), 'Kinds of Knowledge, Limits of Science', in J. de Ridder, H. Peels and R. van Woudenberg (eds), *Scientism: Prospects and Problems*, 190–219, New York: Oxford University Press.

Deane, G. (2021), 'Consciousness in Active Inference: Deep Self-Models, Other Minds, and the Challenge of Psychedelic-Induced Ego-Dissolution', *Neuroscience of Consciousness*, 2 (2), 1–16.

Dembroff, R. and D. Whitcomb (2022), 'Content-Focused Epistemic Injustice', *Oxford Studies in Epistemology* 7, 44–70, Oxford: Oxford University Press.

Dennett, D. C. (2006), *Breaking the Spell: Religion as a Natural Phenomenon*, London: Allen Lane.

Dennett, D. C. (2007), 'Philosophy as Naïve Anthropology: Comment on Bennett and Hacker', in M. Bennett and D. Dennett, P. M. S. Hacker and J. Searle (eds), *Neuroscience & Philosophy*, 73–95, New York: Columbia University Press.

Dennett, D. C. (2010), *Content and Consciousness: Routledge Classics*, London: Routledge.

Deutsch, D. (2010), 'Speaking in Tones', *Scientific American Mind*, July/August: 36–43.
Di Paolo, E. A., E. C. Cuffari and H. De Jaegher (2018), *Linguistic Bodies: The Continuity Between Life and Language*, Cambridge, MA: MIT Press.
Diamond, C. (1995), *The Realistic Spirit: Wittgenstein, Philosophy, and the Mind*, Cambridge, MA: MIT Press.
Diamond, C. (2005), 'Wittgenstein on Religious Belief: The Gulfs Between Us', in D. Z. Phillips and Mario von der Ruhr (eds), *Religion and Wittgenstein's Legacy*, 99–137, London: Ashgate.
Diamond, C. (2019), *Reading Wittgenstein with Anscombe, Going on to Ethics*. Cambridge, MA: Harvard University Press.
Dixson, H. G. W., A. F. Komugabe-Dixson, B. J. Dixson and J. Low (2017), 'Scaling Theory of Mind in a Small-Scale Society: A Case Study from Vanuatu', *Child Development*, 75 (21): 1510, 2157–75.
Dodds, E. R. (1951), *The Greeks and the Irrational*, Berkeley: University of California Press.
Donald, M. (2001), *A Mind So Rare: The Evolution of Human Consciousness*, New York: Norton.
Donald, M. (2004), 'The Definition of Human Nature', in D. Rees and S. Rose (eds), *The New Brain Sciences: Perils and Prospects*, 34–58, Cambridge: Cambridge University Press.
Dormandy, K. (2021a), 'True Faith: Against Doxastic Partiality About Faith (in God and Religious Communities) and in Defence of Evidentialism', *Australasian Philosophical Review*, 5 (1): 4–28.
Dormandy, K. (2021b), 'The Doxastic Norms of Faith: Reply to Commentators', *Australasian Philosophical Review*, 5 (1): 104–15.
Drayson, Z. (2014), 'The Personal/Subpersonal Distinction', *Philosophy Compass*, 9 (5): 338–46.
Driver, J. and C. Spence (2000), 'Multisensory Perception: Beyond Modularity and Convergence', *Current Biology*, 10 (20): R731–R735.
Drury, M. O'C. (2018), *The Selected Writings of Maurice O'Connor Drury*, London: Bloomsbury Academic.
Dunlop, C. E. M. (1984), 'Wittgenstein on Sensation and "Seeing-As"', *Synthese*, 60 (3): 349–67.
Duranti, A. (2008), 'Further Reflections on Reading Other Minds', *Anthropological Quarterly*, 81 (2): 483–94.
Duranti, A. (2015), *The Anthropology of Intentions: Language in a World of Others*, Cambridge: Cambridge University Press.
Engelmann, M. (2013), *Wittgenstein's Philosophical Development: Phenomenology, Grammar, Method, and the Anthropological View*, Basingstoke: Palgrave Macmillan.
Epley, N., B. A. Converse, A. Delbosc, G. A. Monteleone and J. T. Cacioppo (2009), '"Believers" Estimates of God's Beliefs Are More Egocentric than Estimates of Other People's Beliefs', *Proceedings of the National Academy of Sciences*, 106 (51): 21533–8.
Evans, V. (2010), 'The Perceptual Basis of Spatial Representation', in V. Evans and P. Chilton (eds), *Language, Cognition and Space: The State of the Art and New Directions*, 21–48, London: Equinox Publishing.
Fabb, N. (2021), 'Experiences of Ineffable Significance', in E. Ifantidou, L. de Saussure and T. Wharton (eds), *Beyond Meaning*, 324: 135–50, Amsterdam: John Benjamins.
Fauconnier, G. and M. Turner (2002), *The Way We Think: Conceptual Blending and the Mind's Hidden Complexities*, New York: Basic Books.
Fehige, Y., ed. (2016), *Science and Religion: East and West*, Abingdon: Routledge.
Figdor, C. (2018), *Pieces of Mind: The Proper Domain of Psychological Predicates*, Oxford: Oxford University Press.

Figueiredo, F. F. (2015), 'Überreden und Überzeugen. Eine Verhältnisbestimmung aus der Perspektive des Späten Wittgenstein', *Deutsche Zeitschrift für Philosophie*, 63 (1): 58–86.
Figueiredo, F. F. (2019), 'Function as Use. Wittgenstein's Practical Turn in the Early Manuscripts', *Philosophical Investigations*, 42 (1): 66–96.
Figueiredo, F. F. (2021), 'Logischer Holismus und Wittgensteins "Praktische Wende"', *Wittgenstein- Studien*, 12 (1): 195–215.
Fitzgibbon, B. M., M. J. Giummarra, N. Georgiou-Karistianis, P. G. Enticott and J. L. Bradshaw (2010), 'Shared Pain: From Empathy to Synaesthesia', *Neuroscience and Biobehavioral Reviews*, 34 (4): 500–12.
Flavell, J. H. (1976), 'Metacognitive Aspects of Problem Solving', in L. B. Resnick (ed.), *The Nature of Intelligence*, 231–6, Hillsdale: Erlbaum.
Fodor, J. A. (1983), *The Modularity of Mind: An Essay on Faculty Psychology*. Cambridge, MA: MIT Press.
Foley, H. and M. Matlin (2015 [1991]), *Sensation and Perception*, Hove: Psychology Press.
Fraser, R. (2017), 'The Pragmatics and Epistemology of Testimony'. Podcast of Talk at the Moral Sciences Club (Cambridge). Available online: https://sms.csx.cam.ac.uk/media/2481645.
Frazer, J. G. (1890), *The Golden Bough: A Study in Comparative Religion*, vol. 2, London: Macmillan.
Frazer, J. G. (1981), *The Golden Bough*, New York: Random House.
Frege, G. (1893), *Grundgesetze der Arithmetik. Band* I, Jena: Herman Pohle.
Friedman, D. A., A. Tschantz, M. Ramstead, K. Friston and A. Constant (2021), 'Active Inferants: An Active Inference Framework for Ant Colony Behavior', *Frontiers in Behavioral Neuroscience*, 15.
Friedrich, J. and P. Wiener (2020), 'Selection Signatures for High-Altitude Adaptation in Ruminants', *Animal Genetics*, 51 (2): 157–65.
Friston, K. J. (2019), *A Free Energy Principle for a Particular Physics*. https://doi.org/10.48550/arxiv.1906.10184.
Friston, K. J. and C. D. Frith (2015), 'Active Inference, Communication and Hermeneutics', *Cortex*, 68: 129–43.
Friston, K. J., M. Levin, B. Sengupta and G. Pezzulo (2015), 'Knowing One's Place: A Free-Energy Approach to Pattern Regulation', *Journal of the Royal Society. Interface / The Royal Society*, 12 (105).
Friston, K. J., W. Wiese and J. A. Hobson (2020), 'Sentience and the Origins of Consciousness: From Cartesian Duality to Markovian Monism', *Entropy*, 22 (5): 516.
Frith, C. D. and U. Frith (1999), 'Interacting Minds--A Biological Basis', *Science*, 286 (5445): 1692–5.
Frith, C. D. and U. Frith (2006), 'The Neural Basis of Mentalizing', *Neuron*, 50 (4): 531–4.
Frith, U., J. Morton and A. M. Leslie (1991), 'The Cognitive Basis of a Biological Disorder: Autism', *Trends in Neurosciences*, 14 (10): 433–8.
Gallagher, S. (2020), *Action and Interaction*, Oxford: Oxford University Press.
Gardner, H. (1985), *The Mind's New Science: A History of the Cognitive Revolution*, New York: Basic Books.
Geertz, A. W. (2010), 'Brain, Body and Culture: A Biocultural Theory of Religion', *Method and Theory in the Study of Religion*, 22 (4): 304–21.
Geertz, C. (1983), 'Found in Translation: On the Social History of Moral Imagination', in *Local Knowledge: Further Essays in Interpretive Anthropology*, by Clifford Geertz, 36–54, New York: Basic Books, Inc.
Gelfand, M. J., L. H. Nishii and J. L. Raver (2006), 'On the Nature and Importance of Cultural Tightness-Looseness', *The Journal of Applied Psychology*, 91 (6): 1225–44.

Genovese, J. E. C. (2005), 'Paranormal Beliefs, Schizotypy, and Thinking Styles Among Teachers and Future Teachers', *Personality and Individual Differences*, 39 (1): 93–102.

Gerlach, C., J. J. Barton, A. Albonico, M. Malaspina and R. Starrfelt (2022), 'Contrasting Domain-General and Domain-Specific Accounts in Cognitive Neuropsychology: An Outline of a New Approach with Developmental Prosopagnosia as a Case', *Behavior Research Methods*, 54 (6): 2829–42.

Gervais, W. M. (2013), 'Perceiving Minds and Gods: How Mind Perception Enables, Constrains, and Is Triggered by Belief in Gods', *Perspectives on Psychological Science*, 8 (4): 380–94.

Gervais, W. M. and J. Henrich (2010), 'The Zeus Problem: Why Representational Content Biases Cannot Explain Faith in Gods', *Journal of Cognition and Culture*, 10 (3): 383–9.

Gettier, E. (1963), 'Is Knowledge Justified True Belief?', *Analysis*, 23 (6): 121–3.

Glock, H.-J. (1996), *A Wittgenstein Dictionary*, Oxford: Wiley-Blackwell.

Glock, H.-J. (2008), 'Relativism, Commensurability and Translability', in J. Preston (ed.), *Wittgenstein and Reason*, 21–46, Oxford: Wiley Blackwell.

Glock, H.-J. (2020), 'Minds, Brains, and Capacities: Situated Cognition and Neo-Aristotelianism', *Frontiers in Psychology*, 11: 1–14.

Godden, D. and W. H. Brenner (2010), 'Wittgenstein and the Logic of Deep Disagreement', *Cogency: Journal of Reasoning and Argumentation*, 2: 41–80.

Graham, G. (2014), *Wittgenstein & Natural Religion*, Oxford: Oxford University Press.

Graham, J. and J. Haidt (2010), 'Beyond Beliefs: Religions Bind Individuals into Moral Communities', *Personality and Social Psychology Review*, 14 (1): 140–50.

Gray, H. M., K. Gray and D. M. Wegner (2007), 'Dimensions of Mind Perception', *Science*, 315 (5812): 619.

Gray, K., A. C. Jenkins, A. S. Heberlein and D. M. Wegner (2011), 'Distortions of Mind Perception in Psychopathology', *Proceedings of the National Academy of Sciences*, 108 (2): 477–9.

Gray, K., L. Young and A. Waytz (2012), 'Mind Perception Is the Essence of Morality', *Psychological Inquiry*, 23 (2): 101–24.

Grimmel, A. (2015), 'Wittgenstein and the Context of Rationality: Towards a Language-Practical Notion of Rational Reason and Action', *Journal of Language and Politics*, 14 (5): 712–28.

Groark, K. P. (2011), 'Toward a Cultural Phenomenology of Intersubjectivity: The Extended Relational Field of the Tzotzil Maya of Highland Chiapas, Mexico', *Language and Communication*, 33 (3): 278–91.

Grube, D.-M. (2005), 'Justification Rather Than Truth: Gotthold Ephraim Lessing's Defence of Positive Religion in the Ring-Parable', *Bijdragen International Journal of Philosophy and Theology*, 66 (1): 357–78.

Grube, D.-M. (2015), 'Justified Religious Difference: A Constructive Approach to Religious Diversity', *International Journal of Philosophy and Theology*, 76 (5): 419–27.

Grube, D.-M. and W. Van Herck, eds. (2018), *Philosophical Perspectives on Religious Diversity: Bivalent Truth, Tolerance, and Personhood*, London: Routledge.

Guthrie, S. E. (1993), *Faces in the Clouds: A New Theory of Religion*, New York, Oxford: Oxford University Press.

Hacker, P. M. S. (1997), *Wittgenstein*, London: Phoenix.

Hacker, P. M. S. (2013), 'Before the Mereological Fallacy: A Rejoinder to Rom Harré', *Philosophy*, 88 (343): 141–8.

Hacker, P. M. S. (2018), *The Passions: A Study of Human Nature*, Oxford: Wiley-Blackwell.

Hacker, P. M. S. (2019 [1e 1990]), *Wittgenstein: Meaning and Mind (Volume 3 of an Analytical Commentary on the Philosophical Investigations), Part 1: Essays*, Chichester: John Wiley & Sons.

Hall, M. (2011), *Plants as Persons : A Philosophical Botany*, 99–117, Albany: State University of New York Press.
Hanley, P., J. Bennett and M. Ratcliffe (2014), 'The Inter-relationship of Science and Religion: A Typology of Engagement', *International Journal of Science Education*, 36 (7): 1210–29.
Harré, R. (2001), 'Wittgenstein: Science and Religion', *Philosophy*, 76 (2): 211–37.
Harré, R. (2012), 'Behind the Mereological Fallacy', *Philosophy*, 87 (341): 329–52.
Harvey, P. (1993), 'The Mind-Body Relationship in Pali Buddhism: A Philosophical Investigation', *Asian Philosophy*, 3 (1): 29–41.
Heine, S. J. (2001), 'Self as Cultural Product: An Examination of East Asian and North American Selves', *Journal of Personality*, 69 (6): 881–906.
Henrich, J. (2020), *The Weirdest People in the World: How the West Became Psychologically Peculiar and Particularly Prosperous*, New York: Farrar, Straus and Giroux.
Henrich, J., S. J. Heine and A. Norenzayan (2010), 'The Weirdest People in the World?', *Behavioral and Brain Sciences*, 33 (2–3): 61–83.
Herrmann, P., S. R. Waxman and D. L. Medin (2010), 'Anthropocentrism Is Not the First Step in Children's Reasoning About the Natural World', *Proceedings of the National Academy of Sciences*, 107 (22): 9979–84.
Hertzberg, L. (2023), *Wittgenstein on Criteria and Practices*, Cambridge: Cambridge University Press.
Hesp, C., M. Ramstead, A. Constant, P. Badcock, M. Kirchhoff and K. J. Friston (2019), 'A Multi-scale View of the Emergent Complexity of Life: A Free-Energy Proposal', in Georgi Yordanov Georgiev, J. M. Smart, C. L. Flores Martinez and M. E. Price (eds), *Evolution, Development and Complexity: Multiscale Evolutionary Models of Complex Adaptive Systems*, 195–227, Cham: Springer.
Hesp, C., R. Smith, T. Parr, M. Allen, K. J. Friston and M. J. D. Ramstead (2021), 'Deeply Felt Affect: The Emergence of Valence in Deep Active Inference', *Neural Computation*, 33 (2): 398–446.
Heyes, C. (2018), *Cognitive Gadgets: The Cultural Evolution of Thinking*, Cambridge, MA: Harvard University Press.
Heyes, C. and C. D. Frith (2014), 'The Cultural Evolution of Mind Reading', *Science*, 344 (6190): 1243091.
Hick, J. (1988), *God and the Universe of Faiths: Essays in the Philosophy of Religion*, Springer.
Hick, J. (1997), 'The Epistemological Challenge of Religious Pluralism', *Faith and Philosophy*, 14 (3): 277–86.
Hick, J. (2005), *The Interpretation of Religion: Human Responses to the Transcendent*, 2nd edn, New Haven: Yale University Press.
Hintikka, M. B. and J. Hintikka (1986), *Investigating Wittgenstein*, Oxford: Blackwell.
Hipólito, I., M. Baltieri, K. J. Friston and M. J. Ramstead (2021), 'Embodied Skillful Performance: Where the Action Is', *Synthese*, 199 (1): 4457–81.
Hipólito, I. and T. van Es (2022), 'Enactive-Dynamic Social Cognition and Active Inference', *Frontiers in Psychology*, 13.
Hofstede, G. (1983), 'The Cultural Relativity of Organizational Practices and Theories', *Journal of International Business Research*, 14 (2): 75–89.
Hollan, D. (2012), 'Emerging Issues in the Cross-Cultural Study of Empathy', *Emotion Review*, 4 (1): 70–8.
Hood, B., N. L. Gjersoe and P. Bloom (2012), 'Do Children Think That Duplicating the Body Also Duplicates the Mind?', *Cognition*, 125 (3): 466–74.
Hordern, J. (2016), 'Religion and Culture', *Medicine*, 44 (10): 589–92.
Hordern, J. (2020), 'Religion, Culture and Conscience', *Medicine*, 48 (10): 640–3.
Howard-Snyder, D. (2018), 'Three Arguments to Think That Faith Does Not Entail Belief', *Pacific Philosophical Quarterly*, 100 (1): 114–28.

Hoyt, C. (2012), 'Wittgenstein on the Language of Rituals: The Scapegoat Remark Reconsidered', *Religious Studies*, 48 (2): 165–82.
Hutto, D. and E. Myin (2013), *Radicalising Enactivism: Basic Minds Without Content*, Cambridge, MA: MIT Press.
Hutto, D. and E. Myin (2017), *Evolving Enactivism: Basic Minds Meet Content*, Cambridge, MA: MIT Press.
Inglehart, R. (1997), *Modernization and Postmodernization: Cultural, Economic, and Political Change in 43 Societies*, Princeton: Princeton University Press.
James, W. (1977 [1909]), *A Pluralistic Universe*, Cambridge, MA: Harvard University Press.
James, W. (1982), *Varieties of Religious Experience*, London: Penguin Classics.
Jensen, J. (2009), 'Religion as the Unintended Product of Brain Functions in the "standard cognitive science of religion model": On Pascal Boyer, Religion Explained (2001) and Ilkka Pyysiäinen, How Religion Works (2003)', in M. Stausberg (ed.), *Contemporary Theories of Religion: A Critical Companion*, 129–55, London: Routledge.
Jensen, J. (2010), 'Doing It the Other Way Round: Religion as a Basic Case of "Normative Cognition"', *Method and Theory in the Study of Religion*, 22 (4): 322–9.
Jensen, J. (2011), 'Revisiting the Insider-Outsider Debate: Dismantling a Pseudo-Problem in the Study of Religion', *Method and Theory in the Study of Religion*, 23 (1): 29–47.
Jensen, J. (2013), 'Normative Cognition in Culture and Religion', *Journal for the Cognitive Science of Religion*, 1 (1): 47–70.
Jha, V. N. (2017), 'Matter and Consciousness: The Classical Indian Philosophical Approach', in S. Menon, N. Nagaraj and V. V. Binoy (eds), *Self, Culture and Consciousness, Interdisciplinary Convergences on Knowing and Being*, 419–33, Singapore: Springer.
Johnson, D. D. P. (2009), 'The Error of God: Error Management Theory, Religion, and the Evolution of Cooperation', in S. A. Levin (ed.), *Games, Groups, and the Global Good*, 169–80, Dordrecht: Springer.
Kalbe, E., M. Schlegel, A. T. Sack, D. A. Nowak, M. Dafotakis, C. Bangard, M. Brand, S. Shamay-Tsoory, O. A. Onur and J. Kessler (2010), 'Dissociating Cognitive from Affective Theory of Mind: A TMS Study', *Cortex*, 46 (6): 769–80.
Kalén, A., E. Bisagno, L. Musculus, M. Raab, A. Pérez-Ferreirós, A. M. Williams, D. Araújo, M. Lindwall and A. Ivarsson (2021), 'The Role of Domain-Specific and Domain-General Cognitive Functions and Skills in Sports Performance: A Meta-analysis', *Psychological Bulletin*, 147 (12): 1290.
Kapogiannis, D., A. K. Barbey, M. Su, G. Zamboni, F. Krueger and J. Grafman (2009), 'Cognitive and Neural Foundations of Religious Belief', *Proceedings of the National Academy of Sciences*, 106 (12): 4876–81.
Kastel, N. and C. Hesp (2021), 'Ideas Worth Spreading: A Free Energy Proposal for Cumulative Cultural Dynamics', in *Communications in Computer and Information Science*, vol. 1524 CCIS, 784–98, Cham: Springer.
Kavapalu, H. (1995), 'Power and Personhood in Tonga'. *Social Analysis: The International Journal of Social and Cultural Practice*, 37: 15–28.
Keane, W. (2008), '"Others, Other Minds, and Others" Theories of Other Minds: An Afterword on the Psychology and Politics of Opacity Claims', *Anthropological Quarterly*, 81 (2): 473–82.
Keestra, M. (2008), 'The Diverging Force of Imitation: Integrating Cognitive Science and Hermeneutics', *Review of General Psychology*, 12 (2): 127–36.
Keestra, M. and S. J. Cowley (2011), 'Concepts–Not Just Yardsticks, but Also Heuristics: Rebutting Hacker and Bennett', *Language Sciences*, 33 (3): 464–72.
Keil, F. C. and R. A. Wilsonn (2000), *Explanation and Cognition*, Cambridge, MA: MIT Press.

Kelemen, D. (1999), 'Function, Goals and Intention: Children's Teleological Reasoning About Objects', *Trends in Cognitive Sciences*, 3 (12): 461–8.
Kelly, D. and T. Davis (2018), 'Social Norms and Human Normative Psychology', *Social Philosophy and Policy*, 35 (1): 54–76.
Kennedy, J. E., H. Kanthamani and J. Palmer (1994), 'Psychic and Spiritual Experiences, Health, Well-Being, and Meaning in Life', *Journal of Parapsychology*, 58 (4): 353–83.
Kenny, A. J. P. (1971), 'The Homunculus Fallacy', in M. Grene (ed.), *Interpretations of Life and Mind*, 155–65, London: Routledge.
Keyes, C. F. and E. V. Daniel, eds. (1983), *Karma: An Anthropological Inquiry*, Berkeley: University of California Press.
Khambhati, A. N., M. G. Mattar, N. F. Wymbs, S. T. Grafton and D. S. Bassett (2018), 'Beyond Modularity: Fine-Scale Mechanisms and Rules for Brain Network Reconfiguration', *NeuroImage*, 166: 385–99.
Kidwell, C. S., H. Noley and G. E. Tinker (2020), *A Native American Theology*, Maryknoll, NY: Orbis Books.
Kienzler, W. (1997), *Wittgensteins Wende zu Seiner Spätphilosophie 1930 –1932; eine Historische und Systematische Darstellung*, Frankfurt/Main: Suhrkamp.
Kierkegaard, S. (1985 [1843]), *Fear and Trembling*, New York: Viking Penguin.
Kierkegaard, S. (1992 [1846]), *Concluding Unscientific Postscript to 'Philosophical Fragments'*, Princeton: Princeton University Press.
Kirchhoff, M. D. and J. Kiverstein (2020), 'Attuning to the World: The Diachronic Constitution of the Extended Conscious Mind', *Frontiers in Psychology*, 11.
Kitcher, P. (2012), *Preludes to Pragmatism: Toward a Reconstruction of Philosophy*, Oxford: Oxford University Press.
Knitter, P., ed. (2004), *The Myth of Religious Superiority*, Maryknoll: Orbis Books.
Kramer, M. R., P. H. Cox, S. R. Mitroff and D. J. Kravitz (2022), 'A Precise Quantification of How Prior Experience Informs Current Behavior', *Journal of Experimental Psychology: General* 151(8), 1854–1865. (see: https://psycnet.apa.org/record/2022-27231-001).
Kraut, R. (2018), 'Aristotle's Ethics', *Stanford Encyclopedia of Philosophy*. Available online: https://plato.stanford.edu/entries/aristotle-ethics/.
Kripke, S. A. (1982), *Wittgenstein on Rules and Private Language: An Elementary Exposition*, Cambridge, MA: Harvard University Press.
Krull, D. S., M. H.-M. Loy, J. Lin, C.-F. Wang, S. Chen and X. Zhao (1999), 'The Fundamental Fundamental Attribution Error: Correspondence Bias in Individualist and Collectivist Cultures', *Personality and Social Psychology Bulletin*, 25 (10): 1208–19.
Kuchling, F., K. Friston, G. Georgiev and M. Levin (2020), 'Morphogenesis as Bayesian Inference: A Variational Approach to Pattern Formation and Control in Complex Biological Systems', *Physics of Life Reviews*, 33: 88–108.
Kusch, M. (2013), 'Annalisa Coliva on Wittgenstein and Epistemic Relativism', *Philosophia*, 41 (1): 37–49.
Kusch, M. (2020), *Routledge Handbook of Philosophy of Relativism*, London: Routledge.
Kuusela, O. (2014), 'Gordon Baker, Wittgensteinian Philosophical Conceptions and Perspicuous Representation: The Possibility of Multidimensional Logical Descriptions', *Nordic Wittgenstein Review*, 3 (2): 71–98.
Laidlaw, J. (2007), 'A Well-Disposed Social Anthropologist's Problems with the "Cognitive Science of Religion"', in H. Whitehouse and J. Laidlaw (eds), *Religion, Anthropology, and Cognitive Science*, 211–46, Durham: Carolina Academic Press.

Laine, J. W. (2014), *Meta-religion: Religion and Power in World History*, Oakland: University of California Press.
Lakatos, I. (1970), 'Falsification and the Methodology of Scientific Research Programmes', in I. Lakatos and A. Musgrave (eds), *Criticism and the Growth of Knowledge*, 91–196, Cambridge: Cambridge University Press.
Lakoff, G. and M. Johnson (2003), *Metaphors We Live By*, Chicago: University of Chicago Press.
Lasonen-Aarnio, M. (2020), 'Enkrasia or Evidentialism? Learning to Love Mismatch', *Philosophical Studies*, 177 (3): 597–632.
Laurin, K., A. C. Kay and G. M. Fitzsimons (2012), 'Divergent Effects of Activating Thoughts of God on Self-Regulation', *Journal of Personality and Social Psychology*, 102 (1): 4–21.
Lawson, E. T. and R. N. McCauley (1990), *Rethinking Religion: Connecting Cognition & Culture*, Cambridge: Cambridge University Press.
Lebedev, A. V., M. Lövdén, G. Rosenthal, A. Feilding, D. J. Nutt and R. L. Carhart-Harris (2015), 'Finding the Self by Losing the Self: Neural Correlates of Ego-Dissolution Under Psilocybin', *Human Brain Mapping*, 36 (8): 3137–53.
Leslie, A. M. (2001), 'Theory of Mind', in N. Smelser and P. Baltes (eds), *International Encyclopedia of the Social and Behavioral Sciences*, 15652–6, Elsevier.
Lessing, G. E. (2005), *Philosophical and Theological Writings (Cambridge Texts in the History of Philosophy)*, edited by H. P. Nisbet, Cambridge: Cambridge University Press.
Leung, A. K. Y. and D. Cohen (2011), 'Within- and Between-Culture Variation: Individual Differences and the Cultural Logics of Honor, Face, and Dignity Cultures', *Journal of Personality and Social Psychology*, 100 (3): 507–26.
Lewes, G. H. (1891), *The Physical Basis of Mind*, Boston: Houghton Mifflin.
Lillard, A. A. (1998), 'Ethnopsychologies: Cultural Variations in Theories of Mind', *Psychological Bulletin*, 123 (1): 3–32.
Lincoln, B. (2003), *Holy Terrors: Thinking About Religion After September 11*, Chicago: University of Chicago Press.
Lloyd, D. (2006), 'Representation, Philosophical Issues About', in L. Nadel (ed.), *Encyclopedia of Cognitive Science*, 934–40, Hoboken: Wiley.
Locke, J. (1979), *An Essay Concerning Human Understanding*. The Clarendon Edition of the Works of John Locke; Locke, John, 1632–704. Works. Oxford: Clarendon Press.
Löffler, W. (2018), 'Challenges to Moral and Religious Belief: Disagreement and Evolution', *Philosophia Christi*, 20 (1): 306–9.
Loughlin, V. (2021), *4E Cognitive Science and Wittgenstein*, Cham: Palgrave Pivot.
Loy, D. (1982), 'Enlightenment in Buddhism and Advaita Vedanta', *International Philosophical Quarterly*, 22 (1): 65–74.
Luckmann, T. (2019), 'The New and the Old in Religion', in *Social Theory for a Changing Society*, 167–88, London: Routledge.
Luhrmann, T. (2011), 'Toward an Anthropological Theory of Mind', *Suomen Antropologi: Journal of the Finnish Anthropological Society*, 36 (4): 5–69.
Luhrmann, T. M. (2018), 'The Faith Frame: Or, Belief Is Easy, Faith Is Hard', *Contemporary Pragmatism*, 15: 302–18.
Luhrmann, T. M. (2020), *How God Becomes Real: Rekindling the Presence of Invisible Others*, Princeton: Princeton University Press.
Luhrmann, T. M., K. Weisman, F. Aulino, J. D. Brahinsky, J. C. Dulin, V. A. Dzokoto, C. H. Legare, M. Lifshitz, E. Ng, N. Ross-Zehnder and R. E. Smith (2021), 'Sensing the Presence of Gods and Spirits Across Cultures and Faiths', *Proceedings of the National Academy of Sciences*, 118 (5): e2016649118.

Luna, L. E. (1986), *Vegetalismo - Shamanism Among the Mestizo Population of the Peruvian Amazon*, Stockholm: Almqvist & Wiksell International.
Lurie, Y. (2012), *Wittgenstein on the Human Spirit*, Amsterdam: Brill.
Malcolm, N. (1952), 'Knowledge and Belief', *Mind*, 61 (242): 178–89.
Malcolm, N. (1993), *Wittgenstein: A Religious Point of View?*, London: Routledge.
Malle, B. F. (2006), *How the Mind Explains Behavior: Folk Explanations, Meaning, and Social Interaction*, Cambridge, MA: MIT Press.
Malle, B. F. and J. Knobe (1997), 'The Folk Concept of Intentionality', *Journal of Experimental Social Psychology*, 33 (2): 101–21.
Mandler, J. M. (2006), *The Foundations of Mind: Origins of Conceptual Thought*, Oxford: Oxford University Press.
Margolis, J. (1991), *The Truth About Relativism*, Oxford: Blackwell Publishing.
Markus, H. R. and S. Kitayama (1991), 'Culture and the Self: Implications for Cognition, Emotion, and Motivation', *Psychological Review*, 98 (2): 224–53.
Masuzawa, T. (2005), *The Invention of World Religions or, How European Universalism Was Preserved in the Language of Pluralism*, Chicago: University of Chicago Press.
Matheson, J. (2018), 'Disagreement Skepticism and the Rationality of Religious Belief', in Kevin McCain and Ted Poston (eds), *The Mystery of Skepticism*, 83–104, Leiden: Brill.
Matsumoto, D. (1989), 'Cultural Influences on the Perception of Emotion', *Journal of Cross-Cultural Psychology*, 20 (1): 92–105.
Matsumoto, D., S. H. Yoo and J. Fontaine (2008), 'Mapping Expressive Differences Around the World', *Journal of Cross-Cultural Psychology*, 39 (1): 55–74.
Mayeda, S. (1992), *A Thousand Teachings: The Upadeśasāhasrī of Śaṅkara*, Albany: State University of New York Press.
Mayer, A. and B. Trauble (2012), 'Synchrony in the Onset of Mental State Understanding Across Cultures? A Study Among Children in Samoa', *International Journal of Behavioral Development*, 37 (1): 21–8.
mays, b. o., R. Seligman and D. L. Medin (2020), 'Cognition Beyond the Human: Cognitive Psychology and the New Animism', *Ethos*, 48 (1): 50–73.
McCauley, R. N. (1996), 'Explanatory Pluralism and the Coevolution of Theories in Science', in R. McCauley (ed.), *The Churchlands and Their Critics*, 17–47, Oxford: Blackwell Publishing.
McCauley, R. N. (2000), 'The Naturalness of Religion and the Unnaturalness of Science', in Frank C. Keil and Robert A. Wilson (eds), *Explanation and Cognition*, 61–85, Cambridge, MA: MIT Press.
McCauley, R. N. (2011 [pbk edition 2013]), *Why Religion Is Natural and Science Is Not*, New York: Oxford University Press.
McCauley, R. N. (2017), 'Explanatory Pluralism and the Cognitive Science of Religion: Or Why Scholars in Religious Studies Should Stop Worrying About Reduction', in R. McCauley and T. Lawson (eds), *Philosophical Foundations of the Cognitive Science of Religion: A Head Start*, 1–24, London: Bloomsbury Academic.
McCauley, R. N. (2020), 'Recent Trends in the Cognitive Science of Religion: Neuroscience, Religious Experience, and the Confluence of Cognitive and Evolutionary Research', *Zygon*, 55: 97–124.
McCauley, R. N. and E. Cohen (2010), 'Cognitive Science and the Naturalness of Religion', *Philosophy Compass*, 5 (9): 779–92.
McCauley, R. N. and G. Graham (2020), *Hearing Voices and Other Matters of the Mind*, Oxford: Oxford University Press.
McCauley, R. N. and E. T. Lawson (2017), *Philosophical Foundations of the Cognitive Science of Religion: A Head Start*, London: Bloomsbury Publishing.

McGinn, M. (2015), 'Two Senses of "See"', in M. Campbell and M. O'Sullivan (eds), *Wittgenstein and Perception*, 33–44, Abingdon: Routledge.
McGrath, A. E. (2018), *The Territories of Human Reason: Science and Theology in an Age of Multiple Rationalities*, Oxford: Oxford University Press.
McGuinness, B., ed. (2008), *Wittgenstein in Cambridge: Letters and Documents 1911–1951*, Malden: Blackwell Publishing.
McNamara, R. A. (2016), *Morality When the Mind Is Opaque: Intent vs. Outcome Across the Lifespan in Yasawa, Fiji*, thesis, University of British Columbia.
McNamara, R. A. and J. Henrich (2017), 'Kin and Kinship Psychology Both Influence Cooperative Coordination in Yasawa, Fiji', *Evolution and Human Behavior*, 38 (2): 197–207.
McNamara, R. A. and B. G. Purzycki (2020), 'Minds of Gods and Human Cognitive Constraints: Socio-ecological Context Shapes Belief', *Religion, Brain & Behavior*, 36 (1): 1–16.
McNamara, R. A., R. Senanayake, A. K. Willard and J. Henrich (2021), 'God's Mind on Morality', *Evolutionary Human Sciences*, 3 (E6): 1–19.
McNamara, R. A., A. K. Willard, A. Norenzayan and J. Henrich (2019), 'Weighing Outcome vs. Intent Across Societies: How Cultural Models of Mind Shape Moral Reasoning', *Cognition*, 182: 95–108.
Medin, D., S. Waxman, J. Woodring and K. Washinawatok (2010), 'Human-Centeredness Is Not a Universal Feature of Young Children's Reasoning: Culture and Experience Matter When Reasoning About Biological Entities', *Cognitive Development*, 25 (3): 197–207.
Merikle, P. M., D. Smilek and J. D. Eastwood (2001), 'Perception Without Awareness: Perspectives from Cognitive Psychology', *Cognition*, 79 (1–2): 115–34.
Mikolajczak, M., O. Luminet, C. Leroy and E. Roy (2007), 'Psychometric Properties of the Trait Emotional Intelligence Questionnaire: Factor Structure, Reliability, Construct, and Incremental Validity in a French-Speaking Population', *Journal of Personality Assessment*, 88 (3): 338–53.
Miller, G. A. (2003), 'The Cognitive Revolution: A Historical Perspective', *Trends in Cognitive Sciences*, 7 (3): 141–4.
Miller, G. L. (2009), 'Introduction', in J. W. von Goethe, *The Metamorphosis of Plants* [Die Metamorphose der Pflanzen], xv–xxxi, Cambridge, MA: MIT Press.
Millière, R. (2017), 'Looking for the Self: Phenomenology, Neurophysiology and Philosophical Significance of Drug-Induced Ego Dissolution', *Frontiers in Human Neuroscience*, 11: 245.
Monk, R. (1990), *Ludwig Wittgenstein: The Duty of Genius*, London: Jonathan Cape Ltd.
Monk, R. (1991), *Ludwig Wittgenstein: The Duty of Genius*, London: Vintage Book Company.
Mononen, R., M. Niemivirta and J. Korhonen (2022), 'Predicting Mathematical Learning Difficulties Status: The Role of Domain-Specific and Domain-General Skills', *International Electronic Journal of Elementary Education*, 14 (3): 335–52.
Moore, A. C. (1977), *Iconography of Religions: An Introduction*, Philadelphia: Fortress Press.
Moriguchi, Y., T. Ohnishi, R. D. Lane, M. Maeda, T. Mori, K. Nemoto, H. Matsuda and G. Komaki (2006), 'Impaired Self-Awareness and Theory of Mind: An fMRI Study of Mentalizing in Alexithymia', *Neuroimage*, 32 (3): 1472–82.
Mormann, F. and C. Koch (2007), 'Neural Correlates of Consciousness', *Scholarpedia*, 2 (12): 1740.
Moyal-Sharrock, D. (2015), 'Wittgenstein on Forms of Life, Patterns of Life, and Ways of Living', *Nordic Wittgenstein Review*: 21–42.

Moyal-Sharrock, D. (2016), 'The Animal in Epistemology: Wittgenstein's Enactivist Solution to the Problem of Regress', *International Journal for the Study of Skepticism*, 6 (2–3): 97–119.
Moyal-Sharrock, D., ed. (2017), *The Third Wittgenstein: The Post-investigations Works*. Routledge.
Moyal-Sharrock, D. (2021), *Certainty in Action: Wittgenstein on Language, Mind and Epistemology*, London: Bloomsbury Academic.
Mulhall, S. (2015), *The Great Riddle*, Oxford: Oxford University Press.
Mull, M. S. and E. M. Evans (2010), 'Did She Mean to Do It? Acquiring a Folk Theory of Intentionality', *Journal of Experimental Child Psychology*, 107 (3): 207–28.
Mullins-Nelson, J. L., R. T. Salekin and A.-M. R. Leistico (2006), 'Psychopathy, Empathy, and Perspective-Taking Ability in a Community Sample: Implications for the Successful Psychopathy Concept', *International Journal of Forensic Mental Health*, 5 (2): 133–49.
Munz, V. and B. Ritter, eds (2017), *Wittgenstein's Whewell's Court Lectures: Cambridge, 1938–1941: From the Notes by Yorick Smythies*, Oxford: Wiley-Blackwell.
Nair, V. K., T. Rayner, S. Siyambalapitiya and B. Biedermann (2021), 'Domain-General Cognitive Control and Domain-Specific Language Control in Bilingual Aphasia: A Systematic Quantitative Literature Review', *Journal of Neurolinguistics*, 60: 101021.
Nielsen, K. (1967), 'Wittgensteinian Fideism', *Philosophy*, 42 (161): 191–209.
Nielsen, K. (2001), *Naturalism and Religion*, Amherst: Prometheus Press.
Nielsen, K., D. Z. Phillips, N. Bauer, S. Mulhall and B. Szabados (2005), *Wittgensteinian Fideism?*, London: SCM Press.
Nielsen, M., D. Haun, J. Kärtner and C. H. Legare (2017), 'The Persistent Sampling Bias in Developmental Psychology: A Call to Action', *Journal of Experimental Child Psychology*, 162: 31–8.
Norenzayan, A., W. M. Gervais and K. H. Trzesniewski (2012), 'Mentalizing Deficits Constrain Belief in a Personal God', *PLoS ONE*, 7 (5): e36880.
Norenzayan, A., A. F. Shariff, A. K. Willard, E. Slingerland, W. M. Gervais, R. A. McNamara and J. Henrich (2016), 'The Cultural Evolution of Prosocial Religions', *Behavioral and Brain Sciences*, 39 (E1): 1–65.
Norman, D. A. (1993), 'Cognition in the Head and the World: An Introduction to the Special Issue on Situated Action', *Cognitive Science*, 17: 1–6.
Norris, P. and R. Inglehart (2004), *Sacred and Secular: Religion and Politics Worldwide*, Cambridge: Cambridge University Press.
Ojalehto, B. L., D. L. Medin and S. G. García (2017a), 'Conceptualizing Agency: Folkpsychological and Folkcommunicative Perspectives on Plants', *Cognition*, 162: 103–23.
Ojalehto, B. L., D. L. Medin and S. G. García (2017b), 'Grounding Principles for Inferring Agency: Two Cultural Perspectives', *Cognitive Psychology*, 95: 50–78.
Okasha, S. (2018), *Agents and Goals in Evolution*, Oxford: Oxford University Press.
Osto, D. (2018), 'No-Self in Sāṃkhya: A Comparative Look at Classical Sāṃkhya and Theravāda Buddhism', *Philosophy East and West*, 68 (1): 201–22.
Pachis, P. and O. Panagiotidou (2017), 'The Long Way from Cognitive Science to History: To Shorten the Distance and Fill in the Blanks', in L. H. Martin and D. Wiebe (eds), *Religion Explained? The Cognitive Science of Religion After Twenty-Five Years*, 89–95, London: Bloomsbury Publishing.
Palmira, M. (2018), 'Disagreement, Credences, and Outright Belief', *Ratio*, 31 (2): 179–96.

Panagiotidou, O. (2014), 'The Asklepios Cult: Where Brains, Minds, and Bodies Interact with the World, Creating New Realities', *Journal of Cognitive Historiography*, 1 (1): 14–23.

Panagiotidou, O. with R. L. Beck (2017), *The Roman Mithras Cult: A Cognitive Approach*, London: Bloomsbury Publishing.

Panagiotidou, O. (2022), *Healing, Disease and Placebo in Graeco-Roman Asclepius Temples: A Neurocognitive Approach*, London: Equinox Publishing.

Panayotis, P. and O. Panagiotidou (2017), 'The Long Way from Cognitive Science to History: To Shorten the Distance and Fill in the Blanks', in L. Martin and D. Wiebe (eds), *Religion Explained?: The Cognitive Science of Religion After Twenty-Five Years*, 89–96, London: Bloomsbury Academic.

Papineau, D. (2020), 'Naturalism', *The Stanford Encyclopedia of Philosophy* (Summer 2020 Edition), edited by Edward N. Zalta. Available online: https://plato.stanford.edu/archives/sum2020/entries/naturalism/.

Parker, R. (1998), 'Pleasing Things: Reciprocity in Greek Religion', in C. Gill, N. Postlethwaite and R. Seaford (eds), *Reciprocity in Ancient Greece*, 105–25, Oxford: Clarendon.

Pavese, C. (2019), 'The Psychological Reality of Practical Representation', *Philosophical Psychology*, 32 (5): 784–821.

Peacock, J. (2008), 'Suffering in Mind: The Aetiology of Suffering in Early Buddhism', *Contemporary Buddhism*, 9 (2): 209–26.

Pearl, M. S. (2011), '"Militant Modern Atheism" and Religion', *TheKindlyOnes Blog*. Available online: http://thekindlyones.org/2011/10/07/about-militant-modern-atheism-and-religion/ (Accessed 2021).

Peels, R. and René van Woudenberg (2020), *The Cambridge Companion to Common-Sense Philosophy*, Cambridge: Cambridge University Press.

Penelhum, T. (2019), *Reason and Religious Faith*, London: Routledge.

Perloff, M. (1994), 'From Theory to Grammar: Wittgenstein and the Aesthetic of the Ordinary', *New Literary History*, 25 (4): 899–923.

Peters, M. A. (2022), 'Wittgenstein, mysticism and the "religious point of view": "Whereof One Cannot Speak, Thereof One Must Be Silent"', *Educational Philosophy and Theory*, 54 (12): 1952–9.

Peterson, M. and D. Venema (2021), *Biology, Religion, and Philosophy*, Cambridge: Cambridge University Press.

Pezzulo, G., T. Parr and K. J. Friston (2022), 'The Evolution of Brain Architectures for Predictive Coding and Active Inference', *Philosophical Transactions of the Royal Society B*, 377 (1844): 20200531.

Phillips, D. (1993), *Wittgenstein and Religion*, New York: Palgrave Macmillan.

Phillips, D. Z. (2001), *Religion and the Hermeneutics of Contemplation*, Cambridge: Cambridge University Press.

Phillips, D. Z. (2007), 'Philosophy's Radical Pluralism in the House of Intellect - A Reply to Henk Vroom', in Andy F. Sanders (ed.), *Phillips' Contemplative Philosophy of Religion: Questions and Responses*, by D. Z. Phillips, 197–211, Aldershot: Ashgate.

Pietraszewski, D. and A. E. Wertz (2022), 'Why Evolutionary Psychology Should Abandon Modularity', *Perspectives on Psychological Science*, 17 (2): 465–90.

Plant, B. (2011), 'Religion, Relativism, and Wittgenstein's Naturalism', *International Journal of Philosophical Studies*, 19 (2): 177–209.

Plant, B. (2012), 'Philosophical Diversity and Disagreement', *Metaphilosophy*, 43 (5): 567–91.

Plant, B. (2013), 'Wittgenstein, Religious "Passion," and Fundamentalism', *Journal of Religious Ethics*, 41 (2): 280–309.
Plantinga, A. (1967), 'The Existence of God', *The Philosophical Review*, 76 (1): 105–8.
Plantinga, A. (1990), *God and Other Minds: A Study of the Rational Justification of Belief in God*, Ithaca, NY: Cornell University Press.
Plantinga, A. and N. Wolterstorff (1983), *Faith and Rationality: Reason and Belief in God*. Notre Dame, IN: University of Notre Dame Press.
Pojman, L. (1986a), *Religious Belief and the Will*, London: Routledge.
Pojman, L. (1986b), 'Faith Without Belief', *Faith and Philosophy*, 4: 157–76.
Pollack, D., O. Müller and G. Pickel (2012), 'Church and Religion in the Enlarged Europe: Analyses of the Social Significance of Religion in East and West', in Olaf Müller and Detlef Pollack (eds), *The Social Significance of Religion in the Enlarged Europe: Secularization, Individualization, and Pluralization*, 1–26, London: Routledge.
Povinelli, D. J. and S. Giambrone (2001), 'Reasoning About Beliefs: A Human Specialization?', *Child Development*, 72 (3): 691–5.
Powers, A. R., M. Kelley and P. R. Corlett (2016), 'Hallucinations as Top-Down Effects on Perception', *Biological Psychiatry: Cognitive Neuroscience and Neuroimaging*, 1 (5): 393–400.
Prasad, H. S. (2000), 'Dreamless Sleep and Soul: A Controversy Between Vedanta and Buddhism', *Asian Philosophy*, 10 (1): 61–73.
Premack, D. and G. Woodruff (1978), 'Does the Chimpanzee Have a Theory of Mind?', *Behavioral and Brain Sciences*, 1 (4): 515–26.
Purzycki, B. G. (2013), 'The Minds of Gods: A Comparative Study of Supernatural Agency', *Cognition*, 129 (1): 163–79.
Purzycki, B. G., D. N. Finkel, J. Shaver, N. Wales, A. B. Cohen and R. H. Sosis (2012), '"What Does God Know? Supernatural Agents" Access to Socially Strategic and Non-strategic Information', *Cognitive Science*, 36 (5): 846–69.
Purzycki, B. G. and R. A. McNamara (2016), 'An Ecological Theory of Gods' Minds', in H. D. Cruz and R. Nichols (eds), *Advances in Religion, Cognitive Science, and Experimental Philosophy*, 143–69, New York: Continuum.
Purzycki, B. G. and R. Sosis (2010), 'Religious Concepts as Necessary Components of the Adaptive Religious System', in Ulrich Frey (ed.), *The Nature of God: Evolution and Religion*, 37–59, Marburg: Tectum Verlag.
Purzycki, B. G. and R. H. Sosis (2011), 'Our Gods: Variation in Supernatural Minds', in A. C. Elitzur, Laura Mersini-Houghton, Maximilian A. Schlosshauer, Mark P. Silverman, Jack A. Tuszynski, Rüdiger Vaas, H. Dieter Zeh, Ulrich J. Frey, Charlotte Störmer and K. P. Willführ (eds), *Essential Building Blocks of Human Nature*, 77–93, Berlin: Springer.
Putnam, H. (1963), 'Brains and Behavior', in R. J. Butler (ed.), *Analytical Philosophy: Second Series*, 325–41, Oxford: Blackwell Publishing.
Putnam, H. (1978), *Meaning and the Moral Sciences*, London: Routledge and Kegan Paul.
Pyysiäinen, I. (2003), *How Religion Works: Towards a New Cognitive Science of Religion*, Leiden: Brill.
Pyysiäinen, I. (2009), *Supernatural Agents: Why We Believe in Souls, Gods and Buddhas*, New York: Oxford University Press.
Pyysiäinen, I. (2012), 'Cognitive Science of Religion: State-of-the-Art', *Journal of the Cognitive Science of Religion*, 1: 1, 5–28.

Queloz, M. (2021), *The Practical Origin of Ideas: Genealogy as Conceptual Reverse-Engineering*, Oxford: Oxford University Press.
Queloz, M. and D. Cueni (2020), 'Left Wittgensteinianism', *European Journal of Philosophy*, 29 (4): 758–77.
Rad, M. S., A. J. Martingano and J. Ginges (2018), 'Toward a Psychology of Homo sapiens: Making Psychological Science More Representative of the Human Population', *Proceedings of the National Academy of Sciences of the United States of America*, 115 (45): 11401–5.
Radding, C. (1998), 'Crosses, Caves, and Matachinis: Divergent Appropriations of Catholic Discourse in Northwestern New Spain', *The Americas*, 55 (2): 177–203.
Ramstead, M. J. D., C. Hesp, A. Tschantz, R. Smith, A. Constant and K. J. Friston (2021), 'Neural and Phenotypic Representation Under the Free-Energy Principle', *Neuroscience and Biobehavioral Reviews*, 120 (1): 109–22.
Ramstead, M. J. D., A. K. Seth, C. Hesp, L. Sandved-Smith, J. Mago, M. Lifshitz, G. Pagnoni, R. Smith, G. Dumas, A. Lutz, K. Friston and A. Constant (2022), 'From Generative Models to Generative Passages: A Computational Approach to (Neuro) Phenomenology', *Review of Philosophy and Psychology*, 13 (4): 829–57.
Rao, R. P. and D. H. Ballard (1999), 'Predictive Coding in the Visual Cortex: A Functional Interpretation of Some Extra-Classical Receptive-Field Effects', *Nature Neuroscience*, 2 (1): 79–87.
Reber, A. (1993), *Implicit Learning and Tacit Knowledge: An Essay on the Cognitive Unconscious*, New York: Oxford University Press.
Rhees, R. (1965), 'Some Developments in Wittgenstein's View of Ethics', *The Philosophical Review*, 74 (1): 17–26.
Rhees, R., ed. (1981), *Ludwig Wittgenstein, Personal Recollections*, Oxford: Blackwell.
Richardson, A. and T. Uebel, eds (2007), *The Cambridge Companion to Logical Empiricism*, Cambridge: Cambridge University Press.
Ridder, D. D., J. Verplaetse and S. Vanneste (2013), 'The Predictive Brain and the "Free Will" Illusion', *Frontiers in Psychology*.
Riekki, T., M. Lindeman and T. T. Raij (2014), 'Supernatural Believers Attribute More Intentions to Random Movement than Skeptics: An fMRI Study', *Social Neuroscience*, 9 (4): 400–11.
Rival, L. (2012), 'Animism and the Meanings of Life: Reflections from Amazonia', in M. Brightman, V. E. Grotti and O. Ulturgasheva (eds), *Animism in Rainforest and Tundra: Personhood, Animals, Plants and Things in Contemporary Amazonia and Siberia*, 69–81, Oxford, NY: Berghahn Books.
Robbins, J. (2008), 'On Not Knowing Other Minds: Confession, Intention, and Linguistic Exchange in a Papua New Guinea Community', *Anthropological Quarterly*, 81 (2): 421–9.
Robbins, J. and A. Rumsey (2008), 'Introduction: Cultural and Linguistic Anthropology and the Opacity of Other Minds', *Anthropological Quarterly*, 81 (2): 407–20.
Robbins, J., B. B. Schieffelin and A. Vilaça (2014), 'Evangelical Conversion and the Transformation of the Self in Amazonia and Melanesia: Christianity and the Revival of Anthropological Comparison', *Comparative Studies in Society and History*, 56 (03): 559–90.
Ronkin, N. (2005), *Early Buddhist Metaphysics*, London: Routledge.
Ross, L. (1977), 'The Intuitive Psychologist and His Shortcomings: Distortions in the Attribution Process', *Advances in Experimental Social Psychology*, 10: 173–220.
Rowland, R. and R. M. Simpson (2021), 'Epistemic Permissivism and Reasonable Pluralism', in M. Hannon and J. de Ridder (eds), *Routledge Handbook of Political Epistemology*, 113–22, New York: Routledge.

Ryle, G. (1949), *The Concept of Mind*, London: Hutchinson.
Ryle, J. (2010), *My God, My Land: Interwoven Paths of Christianity and Tradition in Fiji*, Farnham: Ashgate.
Safron, A. (2020), 'An Integrated World Modeling Theory (IWMT) of Consciousness: Combining Integrated Information and Global Neuronal Workspace Theories with the Free Energy Principle and Active Inference Framework; Toward Solving the Hard Problem and Characterizing Agentic Causation', *Frontiers in Artificial Intelligence*, 3: 30.
Saito, K., H. Cui, Y. Suzukida, D. E. Dardon, Y. Suzuki, H. Jeong, A. Révész, M. Sugiura and A. Tierney (2022), 'Does Domain-General Auditory Processing Uniquely Explain the Outcomes of Second Language Speech Acquisition, Even Once Cognitive and Demographic Variables Are Accounted For?', *Bilingualism: Language and Cognition*, 25 (5): 856–68.
Saler, B. (2010), 'Theory and Criticism: The Cognitive Science of Religion', *Method and Theory in the Study of Religion*, 22 (4): 330–9.
Samuel, G. (2014), 'Between Buddhism and Science, Between Mind and Body', *Religions*, 5 (3): 560–79.
Sandved-Smith, L., C. Hesp, J. Mattout, K. J. Friston, A. Lutz and M. J. D. Ramstead (2021), 'Towards a Computational Phenomenology of Mental Action: Modelling Meta-awareness and Attentional Control with Deep Parametric Active Inference', *Neuroscience of Consciousness*, 2021 (1): 1–16.
Schaafsma, S. M., D. W. Pfaff, R. P. Spunt and R. Adolphs (2015), 'Deconstructing and Reconstructing Theory of Mind', *Trends in Cognitive Sciences*, 19 (2): 65–72.
Schieffelin, B. B. (2008), 'Speaking Only Your Own Mind: Reflections on Talk, Gossip and Intentionality in Bosavi (PNG)', *Anthropological Quarterly*, 81 (2): 431–41.
Schilbrack, K. (2014), *Philosophy and the Study of Religions: A Manifesto*, Malden: Wiley Blackwell Publishers.
Schilhab, T. (2007), 'Interactional Expertise Through the Looking Glass: A Peek at Mirror Neurons', *Studies in History and Philosophy of Science Part A*, 38 (4): 741–7.
Schjødt, U. (2018), 'Predictive Coding in the Study of Religion: A Believer's Testimony', in *Evolution, Cognition, and the History of Religion: A New Synthesis*, 364–79, Brill.
Schjødt, U. and M. van Elk (2019), 'The Neuroscience of Religion', in *Oxford Handbook of the Cognitive Science of Religion*, 327–48, Oxford: Oxford University Press.
Schjoedt, U., H. Stødkilde-Jørgensen, A. W. Geertz and A. Roepstorff (2009), 'Highly Religious Participants Recruit Areas of Social Cognition in Personal Prayer', *Social Cognitive and Affective Neuroscience*, 4 (2): 199–207.
Schloss, J. P. and M. J. Murray (2011), 'Evolutionary Accounts of Belief in Supernatural Punishment: A Critical Review', *Religion, Brain and Behavior*, 1 (1): 46–99.
Schönbaumsfeld, G. (2007), *A Confusion of the Spheres*, Oxford: Oxford University Press.
Schroeder, S. (2006), *Wittgenstein: The Way out of the Fly-Bottle*, Cambridge: Polity Press.
Schroeder, S. (2020), 'The Emergence of Wittgenstein's Views on Aesthetics in the 1933 Lectures', *Estetika: The European Journal of Aesthetics*, 57 (1): 5–14.
Schweizer, P. (1993), 'Mind/Consciousness Dualism in Sāṅkhya-Yoga Philosophy', *Philosophy and Phenomenological Research*: 845–59.
Schweizer, P. (2019), 'Sāṃkhya-Yoga Philosophy and the Mind-Body Problem', *Prabuddha Bharata*, 124 (1): 232–42.
Schweizer, P. (2020), 'Advaita and the Philosophy of Consciousness Without an Object - CORE Reader', *Prabuddha Bharata*, 125 (1): 156–64.

Schwenkler, J. (2020), 'Risking Belief', in E. Lambert and J. Schwenkler (eds), *Becoming Someone New: Essays on Transformative Experience, Choice, and Change*, 196–211, Oxford: Oxford University Press.
Scott-Kakures, D. (2001), 'High Anxiety: Barnes on What Moves the Unwelcome Believer', *Philosophical Psychology*, 14: 313–26.
Searle, J. (1995), *The Construction of Social Reality*, New York: The Free Press.
Searle, J. (2007), 'Putting Consciousness Back in the Brain', in *Neuroscience and Philosophy*, 97–124, New York: Columbia University Press.
Sebastian, C. L., N. M. G. Fontaine, G. Bird, S. J. Blakemore, S. A. D. Brito, E. J. P. McCrory and E. Viding (2012), 'Neural Processing Associated with Cognitive and Affective Theory of Mind in Adolescents and Adults', *Social Cognitive and Affective Neuroscience*, 7 (1): 53–63.
Sessions, W. L. (1994), *The Concept of Faith: A Philosophical Investigation*, Ithaca: Cornell University Press.
Seth, A. K. (2021), *Being You: A New Science of Consciousness*, London: Penguin.
Seth, A. K., B. J. Baars and D. B. Edelman (2005), 'Criteria for Consciousness in Humans and Other Mammals', *Consciousness and Cognition*, 14 (1): 119–39.
Seth, A. K. and T. Bayne (2022), 'Theories of Consciousness', *Nature Reviews Neuroscience*, 23: 439–52.
Seth, A. K., E. Izhikevich, G. N. Reeke and G. M. Edelman (2006), 'Theories and Measures of Consciousness: An Extended Framework', *Proceedings of the National Academy of Sciences*, 103 (28): 10799–804.
Shiner, R. A. (1974), 'Wittgenstein on the Beautiful, the Good and the Tremendous', *The British Journal of Aesthetics*, 14 (3): 258–71.
Shore, B. (1996), *Culture in Mind: Cognition, Culture, and the Problem of Meaning*, Oxford: Oxford University Press.
Simos, P. (2002), *Vision in the Brain: Organization and Plasticity in the Visual System*, Lisse: Swets and Zeitlinger Publishers.
Singleton, A. (2014), *Religion, Culture and Society: A Global Approach*, London: Sage.
Sinha, C. (2009), 'Objects in a Storied World: Materiality, Narrativity and Normativity', *Journal of Consciousness Studies*, 16 (6–8): 167–90.
Skrzypińska, K. (2021), 'Does Spiritual Intelligence (SI) Exist? A Theoretical Investigation of a Tool Useful for Finding the Meaning of Life', *Journal of Religion and Health*, 60 (1): 500–16.
Smart, N. (1996), *Dimensions of the Sacred: An Anatomy of the World's Beliefs*, Berkeley: University of California Press.
Smart, N. and B. S. Murthy (1996), *East-West Encounters in Philosophy and Religion*, Long Beach, CA: Long Beach Publications.
Smit, H. and P. M. S. Hacker (2014), 'Seven Misconceptions About the Mereological Fallacy: A Compilation for the Perplexed', *Erkenntnis*, 79 (5): 1077–97.
Smith, A. (2006), 'Cognitive Empathy and Emotional Empathy in Human Behavior and Evolution', *The Psychological Record*, 56 (1): 3–21.
Smith, B. (2017), 'Wittgenstein, Naturalism, and Scientism', in Jonathan Beale and Ian James Kidd (eds), *Wittgenstein and Scientism*, 209–24, New York: Routledge.
Smith, B. H. (2009), *Natural Reflections: Human Cognition at the Nexus of Science and Religion*, New Haven, London: Yale University Press.
Smith, D. M. (1998), 'An Athapaskan Way of Knowing: Chipewyan Ontology', *American Ethnologist*, 25 (3): 412–32.

Smith, R., P. Badcock and K. J. Friston (2021), 'Recent Advances in the Application of Predictive Coding and Active Inference Models Within Clinical Neuroscience', *Psychiatry and Clinical Neurosciences*, 75 (1): 3–13.
Sosis, R. H. (2009), 'The Adaptationist-Byproduct Debate on the Evolution of Religion: Five Misunderstandings of the Adaptationist Program', *Journal of Cognition and Culture*, 9 (3): 315–32.
Spengler, O. (1918), *The Decline of the West, Vol. 1: Form and Actuality*, London: George Allen & Unwin.
Sperber, D. (1996), *Explaining Culture: A Naturalistic Approach*, Oxford: Blackwell Publishing.
Sperber, D. and L. Hirschfeld (2007), 'Culture and Modularity', *The Innate Mind: Culture and Cognition*, 2: 149–64.
Spratling, M. W. (2016), 'Predictive Coding as a Model of Cognition', *Cognitive Processing*, 17 (3): 279–305.
Spunt, R. P. and R. Adolphs (2015), 'Folk Explanations of Behavior: A Specialized Use of a Domain-General Mechanism', *Psychological Science*, 26 (6): 724–36.
Sriraman, B. and W. Benesch (2005), 'Consciousness and Science: An Advaita-Vedantic Perspective on the Theology–Science Dialogue', *Theology and Science*, 3 (1): 39–54.
Stanley, J. (2011), *Know How*, Oxford: Oxford University Press.
Stanley, J. and T. Williamson (2017), 'Skill', *Noûs*, 51 (4): 713–26.
Starmans, C. and P. Bloom (2012), 'Windows to the Soul: Children and Adults See the Eyes as the Location of the Self', *Cognition*, 123 (2): 313–8.
Stern, D. G. (1991), 'The "middle Wittgenstein": From Logical Atomism to Practical Holism', *Synthese*, 87: 203–26.
Stern, D. G. (2006), 'How Many Wittgensteins?', in A. Pichler and S. Säätelä (eds), *Wittgenstein: The Philosopher and His Works*, 205–29, Frankfurt/Main: Ontos Verlag.
Szocik, K. and H. Van Eyghen (2021), *Revising Cognitive and Evolutionary Science of Religion*, 87–100, Cham: Springer.
Tate, D. F., M. E. Shenton and E. D. Bigler (2012), 'Introduction to the Brain Imaging and Behavior Special Issue on Neuroimaging Findings in Mild Traumatic Brain Injury', *Brain Imaging and Behavior*, 6 (2): 103–7.
Tattersall, I. (2012), *Masters of the Planet: The Search for Our Human Origins*, New York: Palgrave MacMillan.
Tayal, N. and N. R. Sharma (2020), 'Comparative View of the Eastern and Western Perspectives on the Concept of Ahamkara/Ego', *The International Journal of Indian Psychology*, 8 (3): 570–6.
Taylor, C. (1989), *Sources of the Self: The Making of the Modern Identity*, Cambridge: Cambridge University Press.
Taylor, C. (2007), *A Secular Age*, 1st edn, Cambridge, MA: Belknap Press of Harvard University Press.
Tejedor, C. (2017), 'Scientism as a Threat to Science: Wittgenstein on Self-Subverting Strategies', in Jonathan Beale and Ian James Kidd (eds), *Wittgenstein and Scientism*, 7–27, New York: Routledge.
Thagard, P. (2012), *Introduction to Cognitive Science*, Cambridge, MA: The MIT Press.
Thelen, E. and L. B. Smith (1996), *A Dynamic Systems Approach to the Development of Cognition and Action*, Cambridge, MA: MIT press.
Thomson, R., M. Yuki, T. Talhelm, J. Schug, M. Kito, A. H. Ayanian, J. C. Becker, M. Becker, C. Y. Chiu, H.-S. Choi, C. M. Ferreira, M. Fülöp, P. Gul, A. M. Houghton-Illera, M. Joasoo, J. Jong, C. M. Kavanagh, D. Khutkyy, C. Manzi, U. M. Marcinkowska,

T. L. Milfont, F. Neto, T. von Oertzen, R. Pliskin, A. San Martin, P. Singh and M. L. Visserman (2018), 'Relational Mobility Predicts Social Behaviors in 39 Countries and Is Tied to Historical Farming and Threat', *Proceedings of the National Academy of Sciences*, 115 (29): 7521–6.
Throop, C. J. (2012), 'On the Varieties of Empathic Experience: Tactility, Mental Opacity, and Pain in Yap', *Medical Anthropology Quarterly*, 26 (3): 408–30.
Tillich, P. (1957), *Dynamics of Faith*, New York: Harper and Brothers.
Tillson, J. (2022), 'Rationality, Religious Belief, and Shaping Dispositions: Replies to Carruth, Gatley, Levy, Kotzee and Rocha', *Studies in Philosophy and Education*, 41 (1): 135–49.
Titchener, E. B. (1924), *A Textbook of Psychology*, London: Macmillan.
Tomlinson, M. (2007), 'Mana in Christian Fiji: The Interconversion of Intelligibility and Palpability', *Journal of the American Academy of Religion*, 75 (3): 524–53.
Toren, C. (2004), 'Becoming a Christian in Fiji: An Ethnographic Study of Ontogeny', *Journal of the Royal Anthropological Institute*, 10 (3): 221–40.
Toren, C. (2009), 'Intersubjectivity as Epistemology', *Social Analysis*, 53 (2): 130–46.
Toyoda, Y. (1998), 'To Which Bird Do You Belong?: Totemic Belief Among the Mari, Papua New Guinea', *Senri Ethnological Studies*, 47: 61–77.
Tremlin, T. (2010), *Minds and Gods: The Cognitive Foundations of Religion*, Oxford: Oxford University Press.
Trigg, R. (1973), *Reason and Commitment*, Cambridge: Cambridge University Press.
Trigg, R. (1982), *The Shaping of Man: Philosophical Aspects of Sociobiology*, Oxford: Blackwell.
Trigg, R. (1990), 'Wittgenstein and Social Science', *Royal Institute of Philosophy Supplements*, 28: 209–22.
Trigg, R. (2001), *Understanding Social Science: A Philosophical Introduction to the Social Sciences*, 2nd edn, Oxford: Basil Blackwell.
Trigg, R. (2015), *Beyond Matter: Why Science Needs Metaphysics*, West Conshohocken: Templeton Press.
Trigg, R. (2020), *Monotheism and Religious Diversity*, Cambridge: Cambridge University Press.
Trigg, R. and J. Barrett, eds (2014), *The Roots of Religion: Exploring the Cognitive Science of Religion*, London: Routledge.
Trubody, B. (2019), 'The Structure of Scientific Fraud: The Relationship Between Paradigms and Misconduct', in M. Addis, P. Lane, P. Sozou and F. Gobet, *Scientific Discovery in the Social Sciences*, 67–83, Heidelberg: Springer.
Turner, B. S. (2008), 'Religious Speech: The Ineffable Nature of Religious Communication in the Information Age', *Theory, Culture and Society*, 25 (7–8): 219–35.
Turner, T. S. (2009), 'The Crisis of Late Structuralism: Perspectivism and Animism: Rethinking Culture, Nature, Spirit, and Bodiliness', *Tipití: Journal of the Society for the Anthropology of Lowland South America*, 7 (1): 3–42.
Unsworth, S. J., W. Levin, M. Bang, K. Washinawatok, S. R. Waxman and D. L. Medin (2012), 'Cultural Differences in Children's Ecological Reasoning and Psychological Closeness to Nature: Evidence from Menominee and European American Children', *Journal of Cognition and Culture*, 12 (1–2): 17–29.
Vainio, O.-P. (2020), 'Religious Faith, Certainty, and Doubt', *Kerygma und Dogma*, 66 (3): 248–61.
Van Elk, M. and E. J. Wagenmakers (2017), 'Can the Experimental Study of Religion Be Advanced Using a Bayesian Predictive Framework?', *Religion, Brain and Behavior*, 7 (4): 331–4.

Van Leeuwen, N. (2014), 'Religious Credence Is Not Factual Belief', *Cognition*, 133: 698–715.
Vandello, J. A., D. Cohen and S. Ransom (2008), 'U. S. Southern and Northern Differences in Perceptions of Norms About Aggression: Mechanisms for the Perpetuation of a Culture of Honor', *Journal of Cross-Cultural Psychology*, 39 (2): 162–77.
Varela, F., E. Thompson and E. Rosch (1991), *The Embodied Mind: Cognitive Science and Human Experience*, Cambridge, MA: MIT Press.
Versnel, H. S. (1981), 'Religious Mentality in Ancient Prayer', in H. S. Versnel (ed.), *Faith, Hope and Worhsip. Aspects of Religious Mentality in the Ancient World*, 1–64, Leiden: Brill.
Versnal, H. S. (2011), *Coping With the Gods: Wayward Readings in Greek Theology*, Leiden: Brill.
Vilas, M. G., R. Auksztulewicz and L. Melloni (2021), 'Active Inference as a Computational Framework for Consciousness', *Review of Philosophy and Psychology*: 1–20.
Vilela, H. (2019), *Consciousness in the Vedanta*, thesis, Universidade do Porto.
Vinten, R. (2020), *Wittgenstein and the Social Sciences*, London: Anthem Press.
Virupakshananda, S. (2015), *Sāmkhya Kārikā of Īśvara Kṛṣṇa with the Tattva-Kaumudī of Sri Vācaspati Miśra*, Lulu Press, Inc.
Visala, A. (2011), *Naturalism, Theism and the Cognitive Study of Religion: Religion Explained?*, London: Taylor & Francis.
Waal, F. B. M. de (2008), 'Putting the Altruism Back into Altruism: The Evolution of Empathy', *Annual Review of Psychology*, 59 (1): 279–300.
Walker, D. E., K. A. Smith and E. Vul (2015), 'The "Fundamental Attribution Error" is Rational in an Uncertain World', in *Proceedings of the 37th Annual Meeting of the Cognitive Science Society*, Pasadena, 2547–52.
Watanabe, J. M. (1990), 'From Saints to Shibboleths: Image, Structure, and Identity in Maya Religious Syncretism', *American Ethnologist*, 17 (1): 131–50.
Weed, L. (2006), 'Consciousness in Advaita Vedanta and in Cognitive Science', *Journal of Indian Philosophy and Religion*, 11: 117–38.
Weed, L. E. (2020), 'Why Are Religious Experiences Ineffable? The Question of Mystical-Noetic Knowledge', in *Mysticism and Experience: Twenty-First-Century Approaches*, 149–72, Lanham: Lexington Books.
Weiskopf, D. A. (2020), 'Representing and Coordinating Ethnobiological Knowledge', *Studies in History and Philosophy of Science Part C: Studies in History and Philosophy of Biological and Biomedical Sciences*, 84: 101328.
Whicher, I. (1997), 'The Mind (Citta): Its Nature, Structure and Functioning in Classical Yoga (1)', *Nagoya Studies in Indian Culture and Buddhism*, 18: 35–62.
Whicher, I. (1998), 'The Mind (Citta): Its Nature, Structure and Functioning in Classical Yoga (2)', *Nagoya Studies in Indian Culture and Buddhism*, 19: 23–82.
Whitcomb, D., H. Battaly, J. Baehr and D. Howard-Synder (2017), 'Intellectual Humility: Owning Our Limitations', *Philosophy and Phenomenological Research*, 94 (3): 509–39.
White, C. (2021), *An Introduction to the Cognitive Science of Religion*, London: Routledge.
Whitehouse, H. (2004), *Modes of Religiousity: A Cognitive Theory of Religious Transmission*, Walnut Creek: Altamira Press.
Whyte, C. J., J. Hohwy and R. Smith (2022), 'An Active Inference Model of Conscious Access: How Cognitive Action Selection Reconciles the Results of Report and No-report Paradigms', *Current Research in Neurobiology*, 3: 100036.

Wierzbicka, A. (2006), 'On Folk Conceptions of Mind, Agency and Morality', *Journal of Cognition and Culture*, 6 (1–2): 165–79.
Willard, A. K. and R. A. McNamara (2019), 'The Minds of God(s) and Humans: Differences in Mind Perception in Fiji and North America', *Cognitive Science*, 43 (1).
Willard, A. K. and A. Norenzayan (2013), 'Cognitive Biases Explain Religious Belief, Paranormal Belief, and Belief in Life's Purpose', *Cognition*, 129 (2): 379–91.
Willard, A. K. and A. Norenzayan (2017), '"Spiritual but not religious": Cognition, Schizotypy, and Conversion in Alternative Beliefs', *Cognition*, 165: 137–46.
Williams, B. (1992), 'Left-Wing Wittgenstein, Right-Wing Marx', *Common Knowledge*, 1 (1): 33–42.
Williams, B. (2005), 'Pluralism, Community and Left Wittgensteinianism', in G. Hawthorne (ed.), *In the Beginning Was the Deed: Realism and Moralism in Political Argument*, 29–39, Princeton: Princeton University Press.
Wilson, M. (2010), 'The Re-tooled Mind: How Culture Re-engineers Cognition', *Social, Cognitive, and Affective Neuroscience*, 5 (2–3): 180–7.
Wittgenstein, L. (1922 [1933]), *Tractatus Logico-Philosophicus* trans. by D. Pears and B. F. McGuinness, London: Routledge.
Wittgenstein, L. (1956), *Remarks on the Foundations of Mathematics*, edited by G. H. von Wright, R. Rhees and G. E. M. Anscombe, translated by G. E. M. Anscombe, Oxford: Blackwell Publishing, revised edition 1978.
Wittgenstein, L. (1958), *The Blue and Brown Books*, Oxford: Basil Blackwell.
Wittgenstein, L. (1965), 'A Lecture on Ethics', *The Philosophical Review*, 74: 3–12.
Wittgenstein, L. (1966), *Lectures and Conversations on Aesthetics, Psychology and Religious Belief*, edited by Cyril Barrett, Oxford: Basil Blackwell [2010 edition from Berkeley: University of California Press].
Wittgenstein, L. (1967), *Zettel*, translated by G. E. M. Anscombe, Oxford: Basil Blackwell.
Wittgenstein, L. (1969), *On Certainty*, edited by G. E. M. Anscombe and G. H. von Wright, Oxford: Blackwell Publishing.
Wittgenstein, L. (1970), *Zettel*, edited by G. E. M. Anscombe and G. H. Von Wright, translated by G. E. M. Anscombe, University of California Press.
Wittgenstein, L. (1974), *Philosophical Grammar*, edited by R. Rhees, Oxford: Blackwell Publishing.
Wittgenstein, L. (1975), *Philosophical Remarks*, edited by R. Rhees, Oxford: Blackwell Publishing.
Wittgenstein, L. (1978 [1ème 1956, 2ème 1967, 3e 1978]), in G. E. M. Anscombe and G. H. von Wright (eds), *Remarks on the Foundations of Mathematics*, Oxford: Blackwell Publishing.
Wittgenstein, L. (1979), *Ludwig Wittgenstein and the Vienna Circle: Conversations, Recorded by Friedrich Waismann*, Oxford: Blackwell Publishing.
Wittgenstein, L. (1980a), *Remarks on the Philosophy of Psychology*, edited by G. E. M. Anscombe and G. H. von Wright, translated by G. E. M. Anscombe, vol. 1, Chicago: University of Chicago Press.
Wittgenstein, L. (1980b), *Remarks on the Philosophy of Psychology*, vol. 2, Oxford: Blackwell Publishing.
Wittgenstein, L. (1982a), *Wittgenstein's Lectures, Cambridge, 1932–1935: From the Notes of Alice Ambrose and Margaret Macdonald*, Chicago: University of Chicago Press.
Wittgenstein, L. (1982b), *Last Writings on the Philosophy of Psychology: Preliminary Studies for Part II of Philosophical Investigations, Volume I*, edited by G. E. M. Anscombe, G. H. von Wright and Heikki Nyman, Oxford: Blackwell.

Wittgenstein, L. (1993a), *Philosophical Occasions*, edited by J. Klagge and A. Nordmann, Indianapolis: Hackett.

Wittgenstein, L. (1993b), 'Remarks on Frazer's Golden Bough', in James Klagge and Alfred Norman (eds), *Philosophical Occasions 1912–1951*, 118–55, Indianapolis: Hackett.

Wittgenstein, L. (1998a), *Culture and Value*, rev. edn, edited by G. H. von Wright, Oxford: Blackwell Publishing.

Wittgenstein, L. (1998b), *The Collected Works of Wittgenstein, Electronic Edition*. All Texts Copyright © Blackwell Publishers. Available online: http://www.nlx.com/collections/121 https://www.informationphilosopher.com/tractatus/Collected_Works.pdf (CWW).

Wittgenstein, L. (2003), *Ludwig Wittgenstein: Public and Private Occasions*, edited by J. C. Klagge and A. Nordmann, Lanham: Rowman & Littlefield Publishing Group Publishers.

Wittgenstein, L. (2005), *The Big Typescript, TS. 213*, German-English scholar's edn, edited and translated by C. G. Luckhardt and M. Aue, Malden: Blackwell Publishing.

Wittgenstein, L. (2009 [1e, 1953, 2e 1958, Reprint of English text alone 1963, 3e, 1967b]), *Philosophische Untersuchungen / Philosophical Investigations*, 4th edn, translated by G. E. M. Anscombe, P. M. S. Hacker and Joachim Sculte, edited by P. M. S. Hacker and J. Schulte, Oxford: Wiley-Blackwell.

'Wittgenstein Archives at the University of Bergen' (2016–), *Interactive Dynamic Presentation of Ludwig Wittgenstein's Philosophical Nachlass*. Available online: http://wab.uib.no/transform/wab.php? modus=opsjoner]. Wittgenstein Archives at the University of Bergen under the direction of Alois Pichler, ed., Bergen: Wittgenstein Archives at the University of Bergen.

Woo, B. H. (2014), 'Kierkegaard's Influence on Karl Barth's Early Theology', *Journal of Christian Philosophy*, 18: 197–245.

Wood, C. and J. H. Shaver (2018), 'Religion, Evolution, and the Basis of Institutions: The Institutional Cognition Model of Religion', *Evolutionary Studies in Imaginative Culture*, 2 (2): 1–20.

Woodward, A. L. (2009), '"Infants" Learning About Intentional Action', in A. L. Woodward and A. Needham (eds), *Learning and the Infant Mind*, 227–48, Oxford: Oxford University Press.

Wright, R. (2021), 'Project to Foster Intellectual Humility and to Counter the Psychology of Tribalism'. Templeton Foundation Online. Available online: https://www.templeton.org/grant/project-to-foster-intellectual-humility-and-counter-the-psychology-of-tribalism (Accessed March 2022).

Wright, R. M. (2018), 'The Kuwai Religions of Northern Arawak Speaking Peoples: Initiation, Shamanism, and Nature Religions of the Amazon and Orinoco', *Boletín de Antropología*, 33 (55): 123–50.

Xygalatas, D. and R. McKay (2013), 'Announcing the Journal for the Cognitive Science of Religion', *Journal for the Cognitive Science of Religion*, 1 (1): 1–4.

Zerilli, J. (2019), 'Neural Reuse and the Modularity of Mind: Where to Next for Modularity?', *Biological Theory*, 14 (1): 1–20.

Index

Printed in the USA
CPSIA information can be obtained
at www.ICGtesting.com
LVHW011639241123
764754LV00005B/231